MIXED REALITY IN ARCHITECTURE, DESIGN AND CONSTRUCTION

Mixed Reality in Architecture, Design and Construction

Edited by

XIANGYU WANG
University of Sydney, NSW
Australia

and

MARC AUREL SCHNABEL
University of Sydney, NSW
Australia

Editors
Xiangyu Wang
Design Lab
Faculty of Architecture, Design & Planning
University of Sydney
148 City Road
Sydney NSW 2006
Australia
x.wang@arch.usyd.edu.au

Marc Aurel Schnabel
Architecture & Allied Arts
Faculty of Architecture, Design & Planning
University of Sydney
148 City Road
Sydney NSW 2006
Australia
marcaurel@usyd.edu.au

ISBN: 978-90-481-8067-7 e-ISBN: 978-1-4020-9088-2

Printed on acid-free paper.

9 8 7 6 5 4 3 2 1

springer.com

TABLE OF CONTENTS

PREFACE

Xiangyu Wang and Marc Aurel Schnabel

PREFACE

With the advancement of digital technologies, Architecture, Design and Construction are constantly endeavouring to reach new horizons: Mixed Reality (MR) enriches their processes to these new destinies. In this book, we introduce MR technologies, research and their applications in architecture, design and construction arenas. The chapters address fundamental issues of MR and present impacts they have on these fields. As more researchers progressively employ MR as their base of enquiry, we see a need for a reference guide bringing the existing status of MR into awareness and expanding on recent research.

We include in this book, a range of invited chapters from leading researchers in the field of Mixed Reality in architecture, design and construction. All authors are experts and/or top researchers in their respective areas and each of the chapters has been rigorously reviewed for intellectual content by the editorial team to ensure a high quality. Predominantly, the chapters introduce most recent research projects on theories, applications and solutions of environments that employ MR and its technologies. More specifically, the central focus of the volume is on the manner in which they can be applied to influence practices in architecture, design collaboration, construction and education.

Introduction

To begin, we introduce, define and describe the various realms of Mixed Reality in the chapter *Framing Mixed Realities*. This overview of MR sets the context and frames the scope of the book. We describe the various realms of the Reality-Virtuality continuum and highlight their applications with key research works that are undertaken in the respective areas. This chapter is then followed by four sections in which there are chapters relating to architecture, design collaboration, construction and education.

Mixed Reality in Design Collaboration

Mixed Reality opens new avenues of communication and collaboration between architects, designers and engineers. Hence, researchers have been fascinated by the possibility of MR-mediated design. The four chapters in the second section present current research in this area.

Augmented Virtuality (AV) is explored much less than is AR. As a first trial in using AV in architecture, Xiangyu Wang and Rui Chen present in *Approaches to Augmenting Virtual Design Environments with Reality* an AV-based virtual space for remote design collaboration. Integrated into the AV environment are real-images mapped onto the surface of virtual counterparts, live video streams of participants, and 3D geometries within the environment. He describes the general concept, application scenarios, prototypical implementation, and the use of the AV system in its current state.

The second chapter by Hartmut Seichter *Communication in Augmented Reality Aided Architectural Design* explores the connection between collaborative urban design, Human Computer Interfaces (HCIs) and communication in MR applications. This chapter sheds light on aspects of communication that are particularly interesting in MR applications.

Jin Won Choi wrote the third chapter, *A Technological Review to Develop an AR-Based Design Supporting System.* It reviews the latest in AR technology and shows the manner which it can support various aspects of the construction industry. The author posits the way in which AR can support design and develops a design support system using realistic AR techniques.

The last chapter in this selection is by Xiangyu Wang and Mi Jeong Kim, *Exploring Presence and Performance in Mixed Reality-Based Design Space*, which presents an exploratory study on measuring the extent of presence in an MR-based design space through a comparative study using a tabletop system with two different types of displays: head-mounted displays and 2D screens. This study explores the link between object presence and design performance in immersive MR-based design space while manipulating 3D blocks representing virtual furniture.

As is evident, this chapter envisions a shift of paradigm to AR with the potential for an economic ripple effect, similar to that caused by the appearance of CAD/CAM techniques.

Mixed Reality in Architecture

The third section contains chapters addressing the issues and application of the Mixed Reality concept and technology in architecture. That four chapters address this specific topic in this section, attests to the rising importance with which this issue is regarded in the computer-aided architectural design (CAAD) community.

As Augmented Reality (AR) technology is migrating to mobile phones, it is critical to investigate how this class of technology/devices can be used to support architectural applications. The first chapter by Mark Billinghurst and Anders Henrysson, *Mobile Architectural Augmented Reality*, reviews previous work in the area of mobile AR in architecture, suggests how it could be applied in an architectural setting, and describes promising future research directions.

The second chapter by Bruce Hunter Thomas, *Augmented Reality Visualisation Facilitating the Architectural Process*, examines how a wearable AR computer system can facilitate the architectural design process for the user. He presents an overview of wearable computer technologies and AR and then introduces his AR 'Tinmith' Backpack System computer system.

As a means to digitally archive historical building information, the chapter by Atsuko Kaga, *Simulation of an Historic Building Using a Tablet Mixed Reality System*, proposes the use of 'Tablet MR,' which can overlay on-the-spot photographic images and Virtual Reality (VR) images to realise a simulation for education of a historical building site. Construction of an experimental model, evaluation of accuracy, and suggestion of a system application possibility are also performed.

This section concludes with the chapter by Jules Moloney – *Temporal Context and Concurrent Evaluation*. He explores the question of how MR might be integrated with current practice to enable more considered decision making at the early stages of design. Two ideas are introduced, temporal visualisation and concurrent evaluation, as the conceptual underpinnings of the implementation of MR technology. This chapter also reviews the taxonomy of MR, and clarifies the requirements for a decision support visualisation environment.

Mixed Reality in Construction

The fourth section looks at three applications that stand as exemplars of Mixed Reality in a construction context. They discuss both static and mobile MR-interactions that allow for a new mode of engagement and way-finding in the process of building, construction and related fields.

Phillip S. Dunston and Do Hyoung Shin set out to explore *Key Areas and Issues for Augmented Reality Applications on Construction Sites*. They present potential applications for MR to support construction activities. They study three key areas whereby MR aids the process: inspection coordination and communication.

In his chapter *Tracking Technologies for Outdoor Mixed Reality Applications*, Amin Hammad proposes novel techniques for the way in which virtual models of construction equipment can be operated and viewed by several operators to interactively simulate construction activities on a construction site. He presents a real scenario and discusses a framework for general use.

Finally, Marcus Tönnis and Gudrun Klinker conclude this section with their chapter *Augmented 3D Arrows Reach their Limits in Automotive Environments*. They describe their research in interface and navigation within a 3D MR realm. Based on automotive MR-information support, the research discusses MR as an aid for navigation without distracting the operator. This research has a variety of potential applications in the construction industries, and therefore stands as a sample of one way in which to explore the new possibilities that MR offers to the field.

Mixed Reality in Education and Learning

In the last section, the role of Mixed Reality in learning and education is discussed. Students and educators are not only challenged with teaching and learning to use tools and instruments, they also have to understand the essence of designing. In this capacity, MR can be a medium that allows the exploration and integration of ideas into the domain of the real.

Firstly, Remo Burkhard and Gerhard Schmitt present in their chapter *Visualising Future Cities in the ETH Value Lab: New Methods for Education and Learning* how MR allows not only students, but also urban planners and decision-makers, to engage in research from the very beginning of a design process. They describe how their MR-lab fits into an educational framework and how the facilities are used in city planning and research.

In the following chapter, *Interplay of Domains: New Dimensions of Design Learning in Mixed Realities*, Marc Aurel Schnabel argues that the distance between the idea in the imagination of a design and its representation, communication and realisation can be bridged by employing MR. Especially in an educational context, the reinterpretations of the design ideas in different realms allows students to understand their actions and reflect on their design as a deep learning experience.

In the final chapter, Thomas Kvan discusses in *Debating Opportunities: Learning Design through Different Structures* the role of representation and simulation in the design process. He specifically focuses on the role of the model in MR and as a medium to support learning in design studios. He concludes with a postulation of the contribution MR makes on design learning.

Postscript

To provide the reader with easy access to all content of the book and to provide an overview and guide of relevant literature in the field, we include a Glossary and References from all chapters in the Postscript. Here too, we reflect critically on the investigations presented and propose ways in which the field may evolve in near future. The contributions made in this book are a snapshot of the current research in Mixed Reality that is evolving rapidly opening new horizons to the fields of architecture, design, and construction. The use of MR in these industries will soon become standard like CAAD or Building Information Modelling (BIM). It is exciting to be able to take part in the development of these new possibilities.

Acknowledgements

We express our gratitude to all authors for their enthusiasm to contribute their research as published here. This book would not have been possible without the constructive comments and advice from Professor John Gero, from the Krasnow Institute for Advanced Study. We are also deeply grateful to our external editor Dr Jennifer Gamble at The University of Sydney, whose expertise and commitment were extraordinary. Thanks and appreciation goes to Ms Janie Yip for designing our book cover. We are also grateful to our external reader Ms Mercedes Paulini as well as Ms Rui Wang, whose backup support on things both small and large made the process a pleasant one. Financial aid came from Faculty of Architecture, Design and Planning at the University of Sydney.

[illegible]

[illegible] to all authors for the book [illegible] contributed [illegible] This book could not have been possible without the [illegible] from [illegible] We are [illegible] The [illegible]

[illegible]

1 MIXED REALITIES

Framing Mixed Realities
Marc Aurel Schnabel

FRAMING MIXED REALITIES

MARC AUREL SCHNABEL
The University of Sydney, Australia

Abstract. New 'realities' are emerging. Novel concepts such as Mixed Reality, Augmented Reality and Augmented Virtuality and their supporting technologies influence architecture, design and construction. These realities replace or merge with the normal physical world and they can be tailored to enhance comprehension for specific design and construction activities. The various realms, their research and applications and their relevance to the field are presented and critically reflected upon. Finally the Reality-Virtuality Continuum is analysed regarding its engagement, abstraction and information overlay.

Keywords. Mixed Reality, Reality-Virtuality Continuum, Engagement, Design Environment.

1. Mixed Realities in Architecture, Design and Construction

Various new developments in computing, visualisation and modelling technologies allow Architecture, Engineering, and Construction (AEC) Industries to make use of novel techniques that merge real life situations with computer generated visual information to combine real and virtual spaces (Anders, 2003). Currently, architects, designers and engineers use a variety of instruments to bridge the gap between the idea of a design and its representation hence linking an idea, its communication and realisation. Any tool demands different responses from of the designer, and each instrument introduces different reinterpretations of the design. Subsequently, inherent characteristics and affordances impose a divergence between the idea and its expression.

In this introductory chapter, research findings in design and interaction within realms reaching from reality to virtuality are presented and defined. A variety of realms (Real Reality, Mixed Reality with Amplified Reality, Augmented Reality, Mediated and Diminished Reality, Augmented Virtuality, Virtualised Reality, and Virtual Reality) and their supporting technologies are entering the AEC professions as novel environments for their interactions.

X. Wang and M.A. Schnabel (eds.), Mixed Reality in Architecture, Design and Construction, 3–11.

These 'realities' merge with or replace parts of the physical world. All Mixed Realities share a common philosophy. In a 1931 publication, Husserl and Gibson already discussed how the artificial interacts with the physical world of everyday human activities in order to enrich the experiences of perception, affordance and engagement.

Due to the nature of reality, which is an inherent concept of existence, social-cultural influences redefine technological advancements that alter the understanding of reality. Mixed Reality (MR) is a novelty and often used in highly specific settings; the differences between the definitions of these novel concepts of reality are not clearly identified and their attributes are vaguely defined. Subsequently, some definitions of reality overlap with one another in both concept and implementation. This highlights the need for a structured review addressing effective adoption and settings of these realms and technologies.

Nonetheless, different realities can be tailored to enhance design comprehension and collaboration for specific activities along a design life cycle (Kvan, 2000). A description of the recent research in the field of MR can act as a reference to allow for an effective adoption of MR and its technologies. Since the field is still highly evolving and a variety of research in AEC as well as other fields continues to emerge the following classification can only highlight key aspects by discussing selective and exemplary research projects to define the various realms. Yet it can showcase the benefits of these realms and their used technologies for certain design activities to reveal their implications in design and construction.

MR merges both realms, real and virtual, into a new environment. Virtual Reality (VR) technologies create an intersection wherein real and virtual-world objects are presented together in a single experience. According to Milgram and Colquhoun (1999) the realms, Augmented Reality (AR) and Augmented Virtuality (AV) are the two major subsets lying within the MR range of the Reality-Virtuality (RV) Continuum (Figure 1). AR is an environment where the

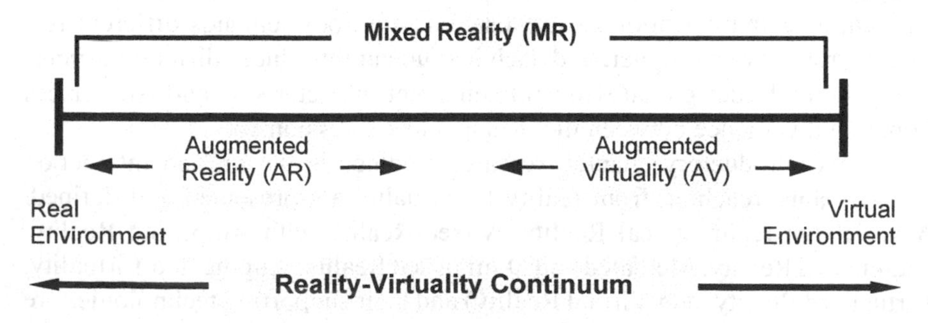

Figure 1. Order of reality concepts ranging from reality (left) to virtuality (right) (adapted from Milgram and Colquhoun, 1999).

additional information generated by a computer is inserted into the user's view of a real world scene. AV, in contrast, creates an environment where real entities are inserted into a predominantly Virtual Environment (VE).

Research in computer-aided MR has been conducted since a couple of decades, but not until the end of the twentieth century, did conferences specialising in MR start to give the field an official platform. These and other related conferences have since established a solid body of international research that is highly recognised in research and industry.

Although the idea to use an MR in AEC is not a novel idea per se, MR has – as presented here in this book – now matured from a purely research field into various practical industrial and consumer applications. Yet a simple straightforward implemented solution for the practice of architecture, design, and construction has yet to be seen.

2. From Reality To Virtuality

As the AEC industries integrate increasingly digitally managed information and Building Information Systems (BIMs), more intuitive visualisation platforms are necessary for efficient use of such information. Recent advances in computer interfaces and hardware instruments have fostered MR prototypes to improve current architectural visualisation, design communication and processes, development of building construction, and engineering management and maintenance systems. The combination of real with virtual entities creates mixed environments that could enhance and aid these processes.

With today's possibilities to influence the RV Continuum, a simple classification such as that presented by Milgram and Colquhoun (1999) is no longer sufficient. Subsequently, it is necessary to incorporate finer subdivisions of the various MRs and to enlarge the scale whilst differentiating between them. Schnabel *et al.* (2007, 2008) summarised their research in MR and VE and established a classification of MR technologies. Figure 2 presents their

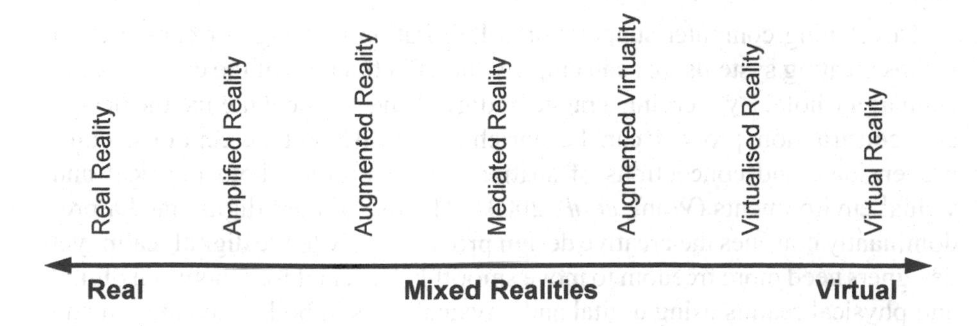

Figure 2. Order of reality concepts ranging from reality (left) to virtuality (right) (Schnabel *et al.*, 2007).

scale of various realities with their degree of reality, where reality is on the left side of the spectrum and virtuality on the right.

In order to frame MR, real and virtual environments need to be defined as well as the various subsets of MRs. The following sections describe these realms within the RV Continuum and highlight research and their applications.

2.1. REALITY

Framing MRs would not be possible without a section about 'Reality' itself. Commonly, reality defines the real and physical world, a realm of elements within this world that actually exist. In the context of architecture and construction, the term reality covers all, that is, whether or not it is created, designed, observable or comprehensible. Reality in this sense may include terms like void, space, solid, building, built, dynamic and stable. There are many philosophical, phenomenological, historical and social categories and definitions of reality that go beyond the framework of this section. It is however interesting to note that, with the emergence of MR, reality redefines itself anew and stands as one realm among others.

2.2. MIXED ENVIRONMENTS

The intersection of real and virtual environments is defined as a Mixed Environment (ME), within which physical and digital elements co-exist, and interact and intermingle in a more expansive form. MR technologies offer potential for interaction between design information and collaborators for the entire life cycle of the engineered facility. Depending on the method of augmentation, ME involves an entire spectrum of environments on the RV Continuum, which are discussed in the following subsections.

An ME, applied in AEC, allows novel ways of collaborative work in shared or remote locations. Examples of applications include collaborative web space (Billinghurst and Kato, 1999) and scientific visualisation (Schmalstieg *et al.*, 2002).

Developing computer support for collaboration in design or construction means creating systems that can amplify the effectiveness of the collaborating team as a whole. By merging a range of digital and physical media, the design and construction process can be enriched by different perceptions, comprehensions, and conceptions of spatial volumes within both physical and virtual environments (Wang *et al.*, 2003). The use of pure digital media predominantly confines the creative design process solely to the digital realm; yet designers need more freedom to move smoothly back and forth between digital and physical realms using digital and physical tools in both conventional and unorthodox ways. The variety of different media transforms the design process from a tangible to a virtual portrayal of architectural design, and vice versa.

As a result of interchanging and crossing-over ranges of design environments from reality to virtuality, the limits of each are dismantled. These realms are subsequently linked together into an overall process leading to alternative form-findings and design-outcomes (Schnabel, 2005).

2.2.1. *Amplified Reality*

To amplify is to enhance properties. Therefore, Amplified Reality means to enrich properties of physical objects with the help of computational means. Falk *et al.* (1999) introduced the concept of Amplified Reality to complement AR, whereby it increases the natural properties of real elements and accentuates the experiences that the objects create within reality. AR is about how the user perceives reality, while Amplified Reality influences how the perceived reality is made available to the user.

Elements within an Amplified Realities embed their properties as parts of themselves. ARs overlay virtual properties onto elements, which in fact do not alter the tangible objects, but rather the perception or experience of it, while elements within an Amplified Reality include their proprietary rights to them. In other words, an amplified object controls the flow of information, and in an AR system the user is in control of that information (Falk *et al.*, 1999).

Projects such as the 'Lovegety' (Iwatani, 1998), 'Hummingbird' (Holmquist *et al.*, 1999) or 'BubbleBadge' (Falk and Björk, 1997) employed an amplified environment. Amplified Reality can be employed in construction and management, for example, to support maintenance systems.

2.2.2. *Augmented Reality*

As a sub-realm of MR based on the definitions by Milgram and Colquhoun (1999), AR adds virtual elements to the perceived reality and allows an interaction in a real world environment while a user receives additional visual computer-generated or modelled information to support the task at hand. In the past, AR environments have been applied primarily in scientific visualisation and gaming entertainment. In recent years, it has been explored for educational and visualisation purposes in AEC and for collaborative design interaction. The virtual elements can be linked with tangible interfaces in order to manipulate them akin to real objects. Though real elements in AR could potentially place certain constraints on the shared imagination, major advantages of AR include ease of collaboration, intuitive interaction, integration of digital information, and mobile computing.

Seichter and Schnabel (2005) argue that this augmentation provides great benefits in architecture and urban design. There is a large variety of research projects that are based on AR in the AEC industries, most of which are based on the 'ARToolkit' technology by Billinghurst and Kato (1999).

In the early phases of a design is the act of sketching, which is a rapid and fuzzy embodiment of the design. The 'sketchand+' system (Seichter, 2003b) is an experimental prototype that makes use of AR in the early stages of architectural design and was found to have a beneficial impact on the quality of the design-process. Broll *et al.* (2004) developed a tabletop AR system for urban planning called 'ARTHUR.' It employs optical see-through displays in connection with a decision-support system. Dunston and Wang's (2005) 'Augmented Reality Computer Aided Drawing' (AR CAD) allows design detailing for mechanical systems, and Yabuki *et al.* (2006) applied AR technology in the erection of steel girder bridges. Often AR is used as collaborative tool that allows multiple users to interact. 'BUILD-IT' by Rauterberg *et al.* (1997) is exemplary as such a multi-user tool that allows planners to engage on a tabletop AR.

2.2.3. *Mediated and Diminished Reality*

Mediated reality describes the general concept of artificial modification of human perception by re-synthesising the light that reaches the eye of a user. Information is added or removed from the scene before it is 'seen' (Starner *et al.*, 1997). Dynamic changes to the appearance and geometry of objects in the real world are typically augmented employing computer graphics and Head Mounted Displays (HMD). Mediated Reality deliberately reduces and alters sensory input. Mediated Reality has been proved to be advantageous in deliberately diminishing the perception of reality.

Lepetit and Berger (2001) describe how objects are deleted from a scene and replaced with suitable backgrounds. For example, this method can aid urban designers to envision a landscape where a building is proposed, removed, or replaced. Bandyopadhyay *et al.* (2001) established Mediated Reality in architectural design in order to alter a building's appearance or to modify sketches. Other applications are found in product design and artwork.

2.2.4. *Augmented Virtuality*

Augmented Virtuality (AV) looks into reality from a virtual world perspective. Milgram and Colquhoun (1999) define AV as the augmentation of a VE with real objects. The AV provides an environment that merges a layered, multimodal, 3D experience into a VE. Despite its potential, Augmented Virtuality has not received as much attention as VR and AR. AV has only been applied in very limited domains such as: displays on unmanned air vehicles (Rackliffe, 2005), 3D video-conferencing systems (Regenbrecht *et al.*, 2004), and a scientific centre (Clarke *et al.*, 2003). According to Oxman (2000) only the commercial design and gaming industries is creating incentives for research in AV.

2.2.5. *Virtualised Reality*

Virtualised Reality communicates reality or scenes of real events by capturing scene descriptions from a number of transcription angles (Kanade *et al.*, 1995). The experience of this realm is a virtual one, whereby the user can adopt any point of view or position within the environment. In other words, Virtualised Reality technology points a set of cameras at a real life event, and allows the viewer to virtually fly around, and to watch the event live or pre-recorded, from completely new positions.

Recently, Virtualised Reality has been used in computer games and there are some developments to employ it as a tool to study building proposals and their usability, such as investigating sight-lines of seating arenas. Kanade (1991) proposed the use of multi-camera stereo using supercomputers for creating 3D models to enrich the virtual world. Rioux *et al.* (1992) outlined a procedure to communicate complete 3D information about an object using depth and reflectance. Fuchs and Neuman (1993) presented a proposal to achieve tele-presence for medical applications.

2.3. VIRTUALITY

On the other side of the RV Continuum is a realm that presents an entirely computer-simulated environment. Computer generated VEs were originally embraced by architects for design concept presentations and visualisations. VR is a constructive tool that aids the designer in the act of designing and communicating within a virtual realm (Davidson and Campbell, 1996). Designers can explore a design without the need of a real artefact. Maze (2002) reports that VEs are mostly used for visualisation of AEC projects and only seldom used for designing itself, such as creation, development, form-finding and collaboration. Likewise Immersive VE (IVE), which enables a certain exclusive degree of immersion of one's senses into a VE, allows active and real-time interactions with virtual design. IVE present new opportunities and answers to design problems through involvement in and with a three-dimensional medium. Schnabel and Kvan (2002) claim that designers are challenged to manage perceptions of solid and void, and navigation and function, without the need to translate to and from physical, mostly two-dimensional media while working in VEs. Subsequently, VEs allows designers to communicate, investigate and express their imagination with greater effortlessness.

In a VE one can change viewpoints and escape gravity, while remaining 'inside' the design without having to convert scales or dimensionalities. Despite the advantages, a translation of design and information from VEs into other realities is potentially problematic. Yet, Yip (2001) found that a re-representation and translation of the design into other environments in fact contributes to the quality of the overall design.

3. Engagement in Architecture, Design and Construction

The definitions of the various realities in the Real-Virtual Continuum allow framing and understanding of the various realms of MEs. The current research of design and construction that employs MR technologies allows for classification of the various realms. They contribute to the understanding of the effects these realities have on the AEC industries. A possible next step is to set up a structure and taxonomy that defines a standard for further research and application.

As MEs were originally embraced for design concept presentations, the advancement of computing allowed designers to interact within the RV Continuum at a more sophisticated level (Hendrickson and Rehak, 1993). In this context it is interesting to study the level of engagement and abstraction MEs offer to the designer and the design. While reality offers a high sensorial engagement, because of the factual existence of the elements, only a low level of abstraction can be experienced. MEs however, offer not only both of these facets, but also additional layers of information, data and other virtual elements that enrich the experience of the user (Figure 3).

Davidson and Campbell (1996) found that MEs and VEs are useful realms to engage in design and communication processes. An ME establishes a co-presence of a shared understanding and knowledge in spatial interactions. Digital models for design or construction are generated with immediacy similar to a physical reality, and constructed to improve the perception and communication of designs. Thus through their high level of interaction and layering of information, MEs provide immediate feedback to its users, which is otherwise impossible within a real or virtual environment. MR allows engaging

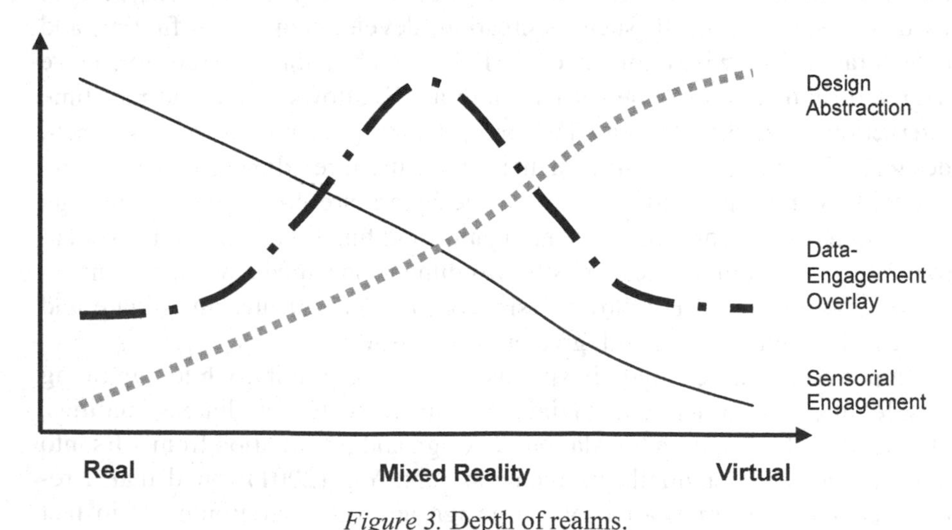

Figure 3. Depth of realms.

through movement and interaction with each and every object and the designers can 'converse' with their designs in multiple realms that overlays additional levels of data.

Yet it is not easy to place designers into an ME. MR technologies and instruments will require further investigation. The presentation format of digital information can be dictated by the features of workspaces. Assumptions about what works and what does not work need to be constantly challenged. Issues such as usability, interface, navigation, clumsiness of gesturing and limited fields of vision constraints have to be further developed to reach the same ease of use and familiarity as real world experiences. A large body of research in MR is investigates questions of usability, however, mostly only in a constrained laboratory environment. Problems with working environments and their tools clearly limit what the designers can do.

This chapter discussed MR realms and their potential advantages over other design environments. Opportunities exist in the early as well as in the final stages of design. Designers can embrace the use of these realms as a medium to converse with designs and each other in novel ways. The potentials for the construction industries are similarly vast. The intersection between real and virtual elements offers an ideal context for a design or construction team to communicate and interact with spatial issues and the handling of complex data and information before, during and after the design process. Clear-cut boundaries between real and virtual are eliminated. It is however, necessary to remind the users of MR that the equipment is only an aid and not a solution or remedy. The role of designers is to be in control of their creativity and their actions. The realities presented here allow a maximum of freedom and a minimum of pre-programmed logic in order to engage rather than restrict creativity. MRs have unlocked new frontiers of engagement in education, collaboration and the profession in architecture and construction.

2 MIXED REALITY IN DESIGN COLLABORATION

APPROACHES TO AUGMENTING VIRTUAL DESIGN ENVIRONMENTS WITH REALITY

XIANGYU WANG AND RUI CHEN
The University of Sydney, Australia

Abstract. This book chapter presents a review of Augmented Virtuality (AV) work and different approaches to realise an AV system for remote design collaboration. Using one of the presented approaches, this chapter also describes an Augmented Virtuality-based virtual space for remote collaboration and inspection. The system allows several participants at different locations to collaborate in an augmented virtual environment simulating a traditional meeting. The general concept, application scenario, prototype implementation, and the use of the AV system in its current state are described in details. This system has the advantage of constructing a virtual environment that incorporates relevant data of the real world into a virtual environment.

Keywords. Augmented Virtuality, Virtual Space, Virtual Reality, Design Collaboration.

1. Introduction

The term Virtual Reality (VR) has been applied in a wide range of situations from the old-fashioned text-based adventure games such as Zork (Lebling *et al.*, 1979) to completely immersive virtual environments such as CAVE (Cave Automatic Virtual Environments). Accordingly, it is therefore unsurprising that the definition of VR varies as well. The commonly held view of a VR environment is one in which the user is completely immersed in a totally synthetic world. Further, this world is understood to mimic some properties of the real world but it also has the capacity to exceed the bounds of physical reality (Milgram *et al.*, 1994). One of the major differences between alternative VR systems is the level of immersion afforded, which could significantly affect the level of presence that users experience through the rendered virtual environment. A low-end example is an ordinary desktop monitor and a high-end example could be head-mounted display (HMD) with a wide field of view.

X. Wang and M.A. Schnabel (eds.), Mixed Reality in Architecture, Design and Construction, 15–26.

Virtual Reality (VR) technology was originally embraced by architecture visionaries for design concept presentations. Over the last decade, computing advances have supported more sophisticated graphics capabilities (Hendrickson and Rehak, 1993).

Presented in this chapter is an Augmented Virtuality-based system that provides the ability for a remote architect to explore a virtual counterpart of a remote place. The virtual counterpart space explored is created to contain real-world images as object textures by mapping certain real elements extracted from the real space onto a virtual environment for richness. The system provides an experience of exploring a virtual representation of a real place. The textures are taken from landmarks/feature objects that exist in the real space. Such texturing creates dual (mirror) objects in the virtual world. This introduces the advantage of making a virtual world appear as the real world and the augmented virtual world could be viewed as a mirrored version of the real place. Another advantage of using real images as textures is to give richness to the virtual environment that contains information and immediate access through visual memory to object identification. Only the critical parts of the real world such as landmarks/defects are necessarily extracted and made salient, and other objects in the real environment are intentionally screened out. The virtual elements look like their real counterparts but can be manipulated in the virtual environment. This makes AV a promising method for viewing visual information from the real world, for example to visually and remotely inspect building construction defects in real time via the virtual environment. Thus, the virtual environment can be used as a control interface/platform for manipulating the corresponding elements from the real world as well as the existing virtual elements.

The most important reason for using augmented virtual environments as control platforms is that those environments can be manipulated and/or navigated in a way that is not subject to certain spatial, temporal, or even physical constraints inherently involved in the real world. For instance, a customised bird's-eye-view or fly-through of the virtual world can be created that is otherwise difficult to realise in a real-world walkthrough. Augmented virtual worlds can be created in an automatic way with textures generated from real world images. The virtual world will have some of the appearance of the real world but maintain the flexibility of the virtual world. Objects can be manipulated in a way that the real world does not allow, for example, objects are not dependent on physical laws, and can be changed according to the needs of the user. In addition, irrelevant parts of the real world can be omitted, to give the user a more easily understood environment without confusing extraneous information.

2. Related Work

Applications in design industries such as architecture, open the door for more innovations in Augmented Virtuality. Despite its potential, Augmented Virtuality has not received nearly the amount of attention paid to Virtual Reality and Augmented Reality. AV has only been applied in very limited domains: such as displays on unmanned air vehicles (Rackliffe, 2005), 3D videoconferencing systems (Regenbrecht *et al.*, 2004), and within a scientific centre (Clarke *et al.*, 2003). There is no recognised research effort in relation to AV applications in the architectural domain. The novelty of the work presented in this chapter is supported by the paucity of published research that investigates AV applications in design and collaboration.

Hughes and Stapleton (2005), who dealt extensively with dynamic real objects, especially in collaborative Augmented Virtuality environments, conducted the first trial of facilitating collaboration in an AV environment. Two users could sit across from each other in a virtual setting; each has a personal point-of-view of a shared virtual environment, and each could also see the other. They used unidirectional retro-reflective material so that each user could extract a dynamic silhouette of the other (Hughes and Stapleton, 2005). These silhouettes could be used to correctly register players relative to each other, and consequently relative to virtual assets.

The Yokohama Character Museum CyberAnnex (Sakagawa *et al.*, 2001a) is a complete virtual representation of an actual museum (The Yokohama Character Museum). It has nine exhibition rooms, and these rooms are described by polygons with texture mapping with approximately 8,000 polygons per room. Images were taken from objects actually being exhibited in the museum, and from which their ray-space data were generated. More than 85 objects described by ray-space data were distributed in three of the rooms. Most of the ray-space data objects in the CyberAnnex were generated with 360 images, and compressed using 18 reference images.

Sakagawa *et al.* (2001b) also developed a scene of the cyber shopping mall where the walls and shelves are described by geometric model data (Figure 1). The toys and flowers are described by ray-space data that was rendered only by the hardware ray-space renderer. The cyber mall has five rooms including a toyshop, a flower shop, and a boutique where the images were taken from real toys, flowers, and clothes.

Augmented Virtuality has also been used in image-guided surgery. For instance, Paul *et al.* (2005) textured 3D preoperative surfaces with camera views and mapped 3D surfaces reconstructed from 2D direct light images of the intra-operative field-of-view to preoperative images. In their work, they presented a system for creating 3D AV scenes for multimodal image-guided neurosurgery. An AV scene includes a 3D surface mesh of the operative field reconstructed from a pair of stereoscopic images acquired through a surgical microscope, and 3D surfaces segmented from preoperative multi-modal images of the patient.

Figure 1. A scene of the cyber shopping mall (Sakagawa *et al.*, 2001b).

Simsarian and Åkesson (1997) developed an AV application for the area of telepresence and tele-exploration. An example of such an application is a security system where a guard virtually roams a domain that is composed of fresh video images taken of the scene. Cameras situated around the space could monitor the security space and apply textures to the virtual model. Thus, the security guard could instead monitor a space by navigating through a virtual world instead of observing 2D video cameras. By adding intelligent cameras and simple image processing, any major changes in the scene could attract the attention of the guard. This should have advantages over 2D video remote monitoring systems.

3. Augmented Virtuality Rendering Techniques

Merging real and virtual entities requires an understanding of the real entities so that they can be located at the right positions and illumination can be correct. There are three major approaches to rendering Augmented Virtuality scenes: video-based, image-based, and model-based. The following sections discuss the techniques that can incorporate real entities into the virtual world. The focus of the discussion is a video-based approach.

3.1. VIDEO-BASED RENDERING TECHNIQUES

There are many reasons for video to be promising in virtual worlds. For example, if a realistic world is preferred, then the use of real video could give the virtual world a more 'natural' appearance. Totally relying on polygons to model a complicated-looking virtual object is not realistic and feasible. The use of video and texture would significantly alleviate the heavy workload and yield a realistic looking model with the potential for real-time updates.

Textures from a video image can be automatically extracted and mapped onto objects. In this way, a virtual place can be constructed from a real place by sweeping the fixed or dynamic camera/s over it. This method is very promising in the areas of telepresence and tele-exploration. An example of such an application, in the context of building science, is a defect inspection system where an architect virtually roams a finished facility, which is composed of fresh video images taken from the real remote site by the local crew. This allows the architect to inspect the potential defects without the need to step out of the office. It is therefore possible to perform parallel inspections for several finished facilities simultaneously, thus maximising the architect's efficiency and expertise.

Augmented Virtuality can create a virtual camera view that can be positioned anywhere inside the environment. Live video can be shown directly in the view frustum and camera-viewing frames are usually depicted by the wireframe view frustum of a camera. As multiple camera interfaces can allow users to more easily understand the situation and better support situational awareness because it can provide scenes of a situation from virtually any perspective. Three of the typical classes of camera perspectives are overhead, third-person, and first-person. The overhead camera perspective can assist a supervisor in gaining a general idea of the situation. A team monitoring the progress of a construction zone would likely use such a perspective frequently. For example, if the task is to obtain additional imagery from a particular construction zone, the overhead perspective can show which parts of the zone have been camera/video-captured. A third-person camera perspective can give close-up views showing more details. For instance, the third-person perspective is useful for remote operators controlling several remote robots to assess the needs of each robot. Operators could quickly judge the relative position and orientation of all the robots. The first-person perspective shows the live video feed of the camera at even closer range.

Video images are embedded into the virtual environment through small 'windows' that are like planes that allow users to look out into the videoed real world. Points projected to the plane/windows through video cameras might be affected by perspective transformation and camera distortion. The rendering of the Augmented Virtuality environment depends on the virtual camera, which contains the intrinsic camera parameters. The pose of the virtual camera is synchronised to one of the real cameras, where the extrinsic camera parameters can be obtained. Camera movement – either of the user's choice or of fixed periodicity – acquires video-textures. Any of these events can trigger the camera to produce textures for each window or plane within its field of view.

The virtual camera compares its own position and one of those planes and then uses that information to extract data from the real camera images. The virtual camera then uses these images as textures. This process can be accomplished by projecting the relevant object surface onto a virtual camera image plane that matches the real camera. As the virtual camera moves around the virtual environment, the virtual world pulls up textures to resemble the real world scene more closely.

3.2. IMAGE-BASED RENDERING TECHNIQUES

Recently image-based modelling and rendering techniques have received much attention as powerful alternatives to traditional geometry-based techniques for image synthesis. Instead of geometric primitives, a collection of sample images is used to render novel views.

Image-based rendering (IBR) uses images, as opposed to polygons, as modelling and rendering primitives. In practice, many IBR approaches correspond to image-geometry hybrids, with the corresponding amount of geometry ranging from per-pixel depth to hundreds of polygons. Image-based modelling (IBM), on the other hand, refers to the use of images to drive the reconstruction of three-dimensional geometric models.

Previous work on image-based rendering reveals a continuum of image-based representations (Lengyel, 1998; Kang, 1999) based on the trade-off between how many input images are needed and how much is known about the scene geometry. These have been classified into three categories of rendering techniques, namely rendering with no geometry, rendering with implicit geometry, and rendering with explicit geometry (Shum and Kang, 2000). These categories should be viewed as a continuum rather than as absolutely discrete, since there are techniques that defy strict categorisation.

The technique used to construct Augmented Virtuality space is IBR, which has been widely adopted in computer graphics. Most image-based rendering techniques have been used in the domains of static environment maps, indoor scenes, or architectural scenes. Examples of IBR include the Light Field (Levoy and Hanrahan, 1996) and Lumigraph (Gortler *et al.*, 1996). The ray-space method (Katayama *et al.*, 1998) uses images of real entities to re-create photorealistic images of complex shaped objects without any explicit geometric model. IBR techniques also have application in Augmented Reality systems. Rendering virtual entities of photorealistic quality is an important precondition to merge seamlessly virtual entities into a real environment (Tamura *et al.*, 1999). IBR uses real image data to render a similar image from an arbitrary perspective. Therefore, the more images that are collected, the more realistic the rendered image looks.

IBR approaches also involve some characteristics that make them less robust. For example, the vast amount of necessary data for IBR requires a large storage space and intensive computing resources to render the images. To achieve an acceptable level of frame rates, the IBR data collected needs to be loaded into the computer memory for fast data access. Accordingly, the size of the computer memory limits the amount of IBR data and compression of the IBR data that is necessary. This is a requirement for more efficient use of memory and, it allows for the potential to increase the quantity of on-memory IBR data available for practical application. In order to offer a more natural-looking virtual scene, texture mapping is an important approach. A texture is typically a bitmap that is mapped onto a surface of an object. There are many ways to map the texture onto the surface of specific 3D objects and texture vertices control the mapping. The standard method for texture-mapping is the same as VRML1.0 (Virtual Reality Modelling Language). VRML1.0 (1995) defines the lower left corner of the bitmap as the origin in the texture coordinate system and the point (1, 1) is the upper right corner. Each vertex of a polygon is given a coordinate in the texture coordinate system, a texture vertex. The texture will be mapped onto the surface so that each texture vertex is mapped to its corresponding surface vertex (VRML1.0 1995). Another issue in warping the texture onto the object is the camera skew. A texture extracted from a camera may not contain right angles.

3.3. MODEL-BASED RENDERING TECHNIQUES

Model-based rendering usually recovers the geometry of the real scene and then renders it from desired virtual view points. Methods for the automatic construction of 3D models have found applications in many fields, including Mobile Robotics, Virtual Reality and Entertainment.

Two categories can be formed from these methods – active and passive. Active methods often require laser technology and structured lights or video, which might result in very expensive equipment. However, new technologies have extended the range of possible applications (Castellani *et al.*, 2002), and new algorithms have improved the ability to cope with problems inherent to laser scanning (Castellani *et al.*, 2002). On the other side, passive methods usually concern the task of generating a 3D model given multiple 2D photographs of a scene. In general they do not require very expensive equipment, but quite often a specialised set-up (Kanade *et al.*, 1995). Passive methods are commonly employed by Model-Based Rendering techniques.

In the model-based rendering technique, Augmented Virtuality requires a model as the basis to construct a view and create a virtual camera perspective. The model can be loosely defined as the collection of all data that has been gathered. For example, a bottom-up perspective of a building structure under construction can be recreated from the model by aligning the virtual camera with that perspective, even if no real camera is actually located. The model can also store historical, extrapolated, and even manually entered data, which can be used to create an Augmented Virtuality environment. One of the main requirements of the model is the ability to store any granularity of data, and display it at the level of detail requested by the user.

One of the advantages of the model-based rendering approach is to alleviate the necessity for the video camera to obtain additional data. For instance, if a user wants detailed data to be collected from a remote monitored place, the user could simply indicate the pre-existing model and obtain data directly via telemetry.

4. Augmented Virtuality System

In this section, we describe the general concept of the Augmented Virtuality system and one envisaged application scenario, which will serve as a case study for comparison of concepts and implementation. The requirements listed are derived from state-of-the-art technology in industry and research. Also presented are the desired properties of the target AV system, future research, and discussions of conceptual alternatives.

4.1. APPLICATION SCENARIO

One application scenario of the Augmented Virtuality system in remote design collaboration is described and, as the field of architectural design is very diverse, the scenario is rather abstract. The AV system provides a communication and collaboration platform for the exchange of a broad range of different types of information. The AV system is embedded in a distributed working environment. Members of a project team situated in different locations need a communication platform to collaborate on their projects. The AV application scenario is that of a planned meeting, not of spontaneous gathering or sharing of information.

In this scenario, two or more engineers/architects are conducting a collaborative design meeting. During this meeting they can discuss changes or improvements to 3D objects derived from CAD data, such as parts, machines, or even entire industrial plants. Since the objects are virtual, the engineers do not have to meet in one location. In addition, either the objects can be discussed as a whole, compared to other objects, or the focus of discussion can be set to details. As in a real meeting, additional documents are necessary for a comprehensive discussion: a slide presentation for the agenda, goals, details and so on; an application sharing capability to discuss details in the CAD system normally used; and text and spreadsheet document display or a possibility for documentation. Each meeting has a coordinator who leads the session and has certain rights and obligations, depending on the purpose of the meeting, such as assigning tasks, choosing or manipulating objects to be displayed, et cetera.

4.2. USER'S WORKPLACE IN THE REAL WORLD

4.2.1. Head-Mounted Display (HMD)

Users can sit at their desks and each wears an HMD audio with headsets. Figure 2 shows the setup situation when a user is interacting with an AV system. The user is equipped with the HMD as the displaying device and controls the orientation through keyboard input/output. The AV system takes in user input such as rotation and reposition, and presents the user with a real-time rendered 3D view displayed on the HMD. The user input is mapped onto the four arrow keys of a standard keyboard. A hi-Res800 HMD is used for the visual feedback to the user. This HMD has two separate displays, which provide 28–24 degree of a diagonal field of view in the horizontal and vertical directions respectively. It also has integrated stereo earphones and an integral microphone, which would be employed for audio conferencing in the AV space. The projection screen displays the first person view from the AV environment and a camera facing the user is placed on the table. This camera captures the video images of the user, which is embedded into the AV environment for communication.

Figure 2. Head-mounted display setup of the AV space (Wang and Gong, 2007).

The main advantage of this setup is the allowance for users to change their view freely inside the augmented environment. Conversely, the main disadvantage is obviously the cumbersome equipment and the limited field of view and resolution. Figure 3 shows an image that could be displayed by the HMD. The AV environment employs both virtual models as well as photos as texture of walls. It clearly shows the detailed environmental view. Corresponding to the user scenario described above, designers are able to perform inspection activities based on this virtual view.

Figure 3. Screenshot of an Augmented Virtuality environment.

4.2.2. *Monitor/Projector*

Systems like FlatWorld, the CAVE (CAVE Automatic Virtual Environment), the NAVE (Non-expensive Automatic Virtual Environment), et cetera have advantages over head-mounted displays (HMDs). Users can sit in front of a standard monitor or stand in front of a multi-screen projection screen (Figure 4)

Figure 4. Multi-screen setup for the AV environment.

that is equipped with one or more cameras, a set of loudspeakers, and a microphone. A special device or metaphor is needed for navigation within the virtual environment such as a sensor pad installed on the floor. The two critical considerations when implementing this setup are creating a sufficient spatial-audio impression with loudspeakers; and tracking the user's head. The simplest and quickest setup would involve the use of a headset to replace loudspeakers, and tracking markers that can be attached to the user's head for tracking.

4.3. INTERACTION WITH THE SYSTEM

The basic interaction is the user's navigation within the environment. The user should be able to see all parts of the AV environment that are important for self-awareness and for the task to be performed. The selection and modification of 3D objects can be realised by the keyboard. The type of object manipulations depends on the task executed, ranging from translation and rotation to assembly or modelling actions. Clear and spatial audio is necessary for effective communication. The spatialisation of the sound should indicate the exact location from which a sound originates. If a participant's virtual avatar is talking in the AV environment, other participants should be able to recognise it even without video.

4.4. VISUALISATION OF AVATARS

To support virtual presence and to provide the impression of 'being there,' graphical/video representations of the participants are needed. In most collaborative virtual reality-based systems, the users are represented either as abstract geometry or as avatars (animated 3D characters). Abstract geometry (like a cone) seldom supports a sense of presence of another person that is acting on the remote site. Avatars are often not convincing unless they have a very high level of detail and behave properly kinematically, which is computationally very expensive.

In AV environments, the participants can be displayed as video images. This provides realistic representations of participants but does not represent the spatial relationship well. The virtual appearance of remote participants should be presented in a way to give sufficient spatially realistic details during distant collaboration (for example, eye contact and eye gaze). For instance, if a remote user turns his/her head, the connected video appearance will move accordingly in the real space. This provides each participant with information about what his or her collaborators are looking at. Even when the user moves freely within the room, the system can follow and trace the movements.

5. Summary

This book chapter gives a thorough review of Augmented Virtuality (AV) work and different approaches to technically realising AV systems. Based on the presented AV approaches, an Augmented Virtuality-based virtual space for remote collaboration and inspection is presented. This chapter introduces the general concept and creation of this Augmented Virtuality prototype that could enhance the intuitive objective of architectural design and collaboration effectiveness by seamlessly inserting real context and experience into a virtual design alternative.

COMMUNICATION IN AUGMENTED REALITY AIDED ARCHITECTURAL DESIGN

Aspects of Collaboration and Communication Using Tangible User Interfaces in Augmented Reality Aided Design (ARAD)

HARTMUT SEICHTER
Human Interface Technology Laboratory, New Zealand

Abstract. This chapter explores the connection between collaborative urban design, HCI (human computer interaction) and communication in Mixed Reality (MR) or Augmented Reality (AR) applications. Due to its nature, architectural design is a joint effort and therefore involves more than one stakeholder. A large portion of the design process is communication and potentially benefits from digital design tools. This chapter sheds light on aspects of communication that are particularly interesting in mixed and augmented reality applications. Because the virtual environment blends in with the real environment, communication facets such as gaze awareness, social presence and other human factors help to provide a framework on which the use of such media can be evaluated. The goal is to outline a new field of design research which targets the mediated interrelationship of real and virtual space introduced through AR and MR technologies.

Keywords. Augmented Reality, Collaboration, Communication.

1. Introduction

Digital design tools are omnipresent in design practice and have helped architects both past and present to explore the functional and formal solution space for architectural problems. Consequently, these digital aids span from dabbling to construction and are already beyond the constraints of pen and paper or other conventional media. Digital design tools were predominantly developed by fostering the capabilities of conventional tools in such a way that they appear as a logical enhancement from their predecessors. However, this legacy also introduces a constraint that is inherent to the physical nature of

X. Wang and M.A. Schnabel (eds.), Mixed Reality in Architecture, Design and Construction, 27–51.

conventional design tools. The objects to be designed can either be virtual or real. With the use of MR and AR in particular the question arises how both virtual and real can co-exist in a meaningful way.

An initial milestone in the research of collaborative digital design tools was created by Bradford *et al.* (1994) coining the term the Virtual Design Studio and identifying means of communication as a key factor for design within new media. Hence, research about digital design started to focus on probing and observing of different media provided by emerging interface technology. Hirschberg *et al.* (1999) looked at the utilisation of time zones for a virtual design process and investigated the impact of the time shifts on the design process. Again, design communication was paramount as the focal point for this investigation. Schnabel and Kvan (2002) extended this work by observing the quality of design and analysed communication across different settings, like Immersive Virtual Environment (IVE) and desktop VE. Gao and Kvan (2004) measured the influence of a communication medium in dyadic settings based on protocol analysis. Another study by Kuan and Kvan (2005) investigated the influence of voxel-based desktop 3D modelling between two different technical implementations. Another important step was to extend the aforementioned body of work into the domain of Augmented Reality and Tangible User Interfaces (TUI).

In order to investigate the impact of TUI and AR an evaluation experiment, integrated into an urban design studio was used. This helped to observe and measure the usage of AR in a practice-near experience for the user and provides real world relevant data. This directly contributes new and transferable knowledge regarding the impact of different affordances of creation interfaces for usage within the design process.

Few studies present methodologies relating to usability evaluation of AR. There is an immediate need to provide these methodologies, and architectural design is not only helpful in this regard, it also is a fertile ground for user evaluation. Future research in this area will lead to applicable guidelines and a framework for the development of simulation environments beyond architecture.

A prominent body of research is the work at Virginia Tech by Hinckley *et al.* (1997) and Gabbard *et al.* (1999) most of which focused on VEs. However the taxonomies for usability evaluation are easily transferable to AR and architectural design. Bowman *et al.* (2002) point out that usability frameworks revolve around specific user groups with group specific 'problems' at hand. Inherently usability evaluations are tailored to a task.

The experiment we introduce here follows this realm with focus on architectural design and utilises a typical setting in a design studio where designers need to comprehend a spatial problem in an early stage of the design process. Comparable to this study is the one presented by Tang *et al.* (2003)

with the aim of assessing the feasibility of augmented assembly. However, it did not take into account the actual communication issues important in a design setting where designers and their peers collaborate. Billinghurst *et al.* (2003) introduced a methodology covering a wider angle of aspects in relation to AR and communication. However, their experiment did not assess the factor of presence.

A usability evaluation provides insight into how users perceive and interact with the system in a praxis-relevant setting. It permits formulation of hypotheses about the relationships between interface, communication and the design task. It therefore is a valuable tool to gauge the impact of human factors in the design process. However, a usability evaluation will not provide an absolute measurement rather; it reveals a tendency for improvement or degradation compared to other interfaces. Furthermore, a user assessment regarding a design creation tool is not free from cultural and educational influences. Therefore the data presented in this section might only be valid within the particular user population in which the experiment was conducted.

The aim of this study was to investigate TUI, which lend themselves as a common ground for discussion in comparison to digitizer pens, which are inherently single user interfaces and have been adapted from 2D for the use in 3D. By using an urban design project, one can tap into a viable scenario for a collaborative AR system as it usually involves large site models, which need to be accessed by multiple users. Physical urban design models can be large in size and therefore difficult to handle easily. The physical properties also limit the ability of the model to present morphological information within its spatial context.

For an assessment of the communication pattern, the chosen scenario is valuable because urban design models are shared with several parties in order to discuss, analysis and re-represent design. These actions require the ability to change and amend parts *in-situ* and visualise and discuss their impact. The experimentation utilised a formal investigation task in order to evaluate the user input devices. This was preferred to the use urban morphology methods, because methodological preferences would have masked the actual impact of the user interface.

2. Experiment

The design of the experiment used two conditions, which only differed in the affordance (Gibson, 1979; Norman, 1988) of the input device for a manipulation task. The objective was to gauge differences between affordances in relation to communication. One setting used an indirect manipulating pen-like interface and the other a direct manipulating TUI (Tangible User Interface).

2.1. CONDITIONS

A pen is an omnipresent tool in general, and specifically in architectural design. As a user interface, the pen is a single-user two-dimensional device which is pressed against a surface in order to create a stroke. The affordances of pressure and the thickness of the line are linked, thus, physical properties of shape, weight, grip and texture provide the user with cues about its handling. Nevertheless, tools like digitizing boards use this perceived affordance to overlay it with other interactions like mouse movements or graffiti spray. Therefore the pen represents a time-multiplexed (Fitzmaurice and Buxton, 1997) interface that relies on a secondary constraining entity. The pen used in this experiment is a 3D input device, which mimics a real pen but uses a button in order to trigger an action rather than being pressed onto a surface.

Building cubes inherently afford a shape which can be stacked and used concurrently. Lego bricks for instance are also known to most of us and the building block is the most basic of manual construction materials. The cube can embody the tool and the object together. The cube interface is therefore a space-multiplexed input device (Fitzmaurice *et al.*, 1995) that affords physical engagement with the object itself.

The experimental set-up measured the differences between input devices regarding factors like presence, communication patterns and performance. Presence is a substantial factor for measuring effects on task performance in VE (Nash *et al.*, 2000). This methodology allows guidelines to be developed for communication enhancing techniques using AR aided collaborative architectural design tools.

2.2. METHOD

This section will introduce the experimental method used and also contains a detailed description of the objective of the design brief given to the participants. It covers the two conditions that were compared, a conventional 3D pen and a tangible user interface. The section continues with an explanation of the conduct of the experimental procedures. Because human factors were a major concern for the experiment, the heuristics in the pilot tests and the ethics concerning simulation environments will be highlighted.

2.2.1. Interfaces

The experimental design was comprised of two phases, an observation phase and a creation phase. The observation phase utilised a direct manipulated interface (AR Toolkit Markers) to view design proposals and in the creative phase, users were asked to extend the design proposal. The variables in this experiment were the input interfaces for the creation phase of the design task.

The observation phase served as a priming phase. One condition implemented the second phase, or creative phase, through directly manipulated cubes (i.e. AR superimposed polystyrene cubes), the other through a tool-object metaphor with a 3D pen based interface (i.e. Polhemus Isotrak with Stylus). In both conditions the users built an abstract massing model representing the extension consisting of a fixed amount of cubes.

2.2.2. *Pen Condition*

The pen based condition utilised the Polhemus Isotrak magnetic tracking system with a Stylus input sensor attached. The Stylus' shape imitates a conventional drawing pen. An initial version, which was used in the pilot tests, used a Smartech SmartBoard. Unfortunately, this device proved to be unreliable in the way that it was used and needed to be replaced. The Polhemus Stylus pen partly mimics the affordance of a real pen but it requires the user to press a button in order to trigger an action. It does not, unlike the SmartBoard and other pen-based technologies react to pressure. The pen interface differed from the other condition as the 3D pen allowed the users to create objects in empty space or within other objects.

2.2.3. *Cube Condition*

For the cube condition, physical cubes representing building bricks, were augmented with a virtual counterpart. Not only does this represent a TUI it also avoids visual registration problems with real objects. In a conventional AR, setting real objects would be occluded by the video overlay. To circumvent this, cubes were re-represented by a virtual pendant. The result was that the graphics engine renders the virtual cubes with the correct occlusion. The pilot test of this condition was conducted with additional props for observation, which were subsequently removed in order to make both conditions comparable based on just one variable. The cube interface afforded stacking of cubes so that the objects could physically interact.

3. Apparatus, Software and Furniture

An important aspect of this research was to create a reliable digital design tool facilitating urban design and providing a practical means for its evaluation. It is a continuation of research about the nature and the impact of immersive digital design tools on the design process. This section briefly outlines the design and development of the software prototype. This process required the design and testing of several subsystems on which the application could be built.

One approach to implement a prototype, which is used in a user evaluation, is to collect and integrate a variety of development tools and libraries. The advantage is that one can draw from features of existing components. Over the years the open-source community yielded a huge collection of utility libraries to create software tools, including for the domains of AR, VR and simulation. Implementation using these existing components can use all features inherited from the components. However, this also introduces the problem that development libraries have diverse target audiences and implementation strategies, and tend to cover a wide variety of functions. Therefore, a large part of development needs to be dedicated to integration and data conversion between the components with all the adaptations for the specific needs of the research intent.

The implementation of a prototype in this particular case goes beyond the collective mechanics of the tool. This is crucial because the intention is to find/create communication artefacts with AR rather than to introduce frustration with the shortcomings of the tool. Therefore, this section outlines hardware issues and those regarding interface design and implementation.

The coherent integration of existing tools to implement specific tasks can become complex. A benefit of this approach can only be expected in the breadth of functionality rather than fast implementation. This is partly an issue of the necessary explicit adaptation. Dependencies of various libraries and tools can lead to a sturdy technological framework, which only can be deployed with considerable effort. The projection and superimposition of specific needs onto a generalised API can itself become a complex task that results in an inflexible tool. Thus, the academic impulse to implement a new toolkit originated because of the lack of a simplified library for the domain of immersive design software (with a very basic but modern functionality that affords adaptations with very short turnaround time and easy deployment to test on diverse hardware). The problem was not the functionality provided by existing toolkits, rather it was their inflexibility and their divergent approaches to implementation. Thus, a RAD tool for this niche audience of digital immersive design started as the TAP (The Architectural Playground) framework in 1999 and is now freely available under an open source license (SourceForge.net, 1999).

Other AR authoring tools and libraries exist but are focused on other domains. In general, a conceptual difference exists between content design and content programming. Tools like AMIRE (Grimm *et al.*, 2002) and DART (MacIntyre *et al.*, 2004) are high-level content design tools. That means they provide capabilities to create AR content without programming. OSGAR (Coelho *et al.*, 2004) and Studierstube (Schmalstieg *et al.*, 2002) are libraries providing users with the possibility of high level programming in order to create content. The TAP framework is a content programming toolkit with a system in place, called ScripTAP, for low-level content design.

3.1. FURNITURE

A table is an omnipresent collaborative entity. In order to support the use of conventional models, the existing system was changed to replace the touch sensitive SmartBoard with a conventional magnetic tracker. Modularisation of the system helped to divide the components and to create furniture supporting rugged use within a design studio. The main design constraint for the hardware (and partly for the software as well) was that it needed to have the capacity be able to be quickly set it up and torn down. Cables were kept as short and unobtrusive as possible. All computers needed to start up and shut down using a single switch including the HMDs and the tracking system. To support applications beyond the experimentation conducted for this study, the system also provided an accessible web interface that when started could load different models.

3.2. APPARATUS

The 'BenchWorks' hardware ran on standard PCs for compatibility and for reasons of easy maintenance. The client systems consisted of two identical Dell Optiplex GX260 systems with 2.8 GHz Intel Pentium 4 HT CPU, 512 MB of RAM, Western Digital 80 GB hard drives, Intel Pro 1,000 MT Adapter and nVIDIA Geforce FX 5200 graphics adapter with 128 MB of DDR2 VRAM with AGP8x enabled. The video capturing was carried out with two identical Philips ToUCam Pro IIs (840 K) with USB 2.0 HighSpeed interface. The server connecting the client systems and handling the start-up sequence was a generic 1.5GHz Pentium 4 system with 512 MB RAM and an Intel Pro 10/100 Ethernet Adapter. All systems were interconnected with the local LAN at The University of Hong Kong through a separate SMC 4-port switch in order to reduce interference with network traffic.

The HMDs used in this experiment were of different types. One was an Olympus FMD-700; the other was an i-glasses SVGA Pro. The Olympus Eye-Trek was connected through the S-Video interface with an AVERmedia Video splitter, the i-glasses HMD was connected directly through a DVI-VGA adapter. There were no additions made in order to adjust the HMDs or lighten them for the user.

The initial input system was a SmartTech SmartBoard which provided a sensitive surface holding the multimarker pattern for AR Toolkit. Due to technical problems it was replaced for the final experiment with a Polhemus Isotrak with a Stylus attached. The actual size of the table top area is based on the original SmartBoard; therefore the dimensions are 152 × 148 cm for the table top with a working height of 90 cm (Figure 1).

Figure 1. Benchwork visualisation from inside.

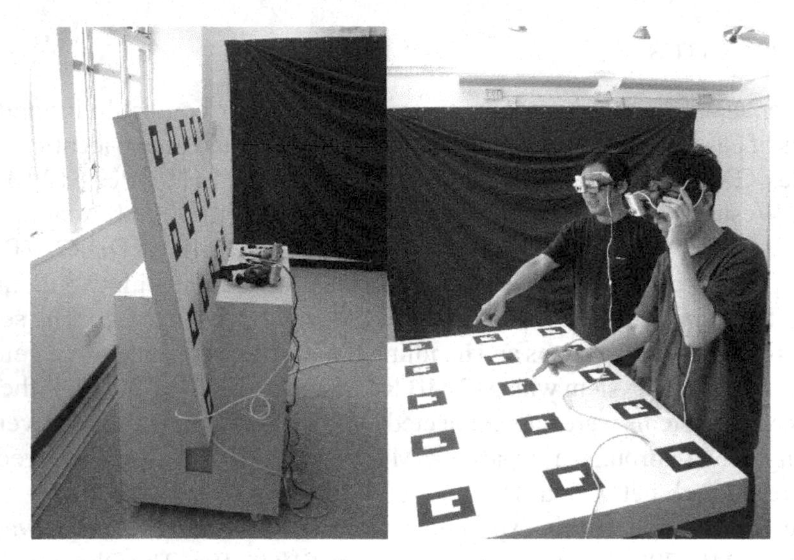

Figure 2. Left: BenchWorks AR table in parked position; right: two users discussing a proposal.

3.3. SOFTWARE ENVIRONMENT

The client computers of the BenchWorks systems used Windows XP Professional with Service Pack 2 applied. To date, the Windows platform is the most appropriate platform for video capture with USB webcams and is therefore preferred for AR applications. No special performance tuning in the software or the operating system was made. Hence, the internal firewall of Windows XP and the other security measures on the client and server machines were switched on through the trials.

The complete configuration of the system was started from a network mounted disk drive in order to share the software and all settings. The software packages and all involved computers were started through an XML-RPC enabled mechanism (XML-RPC.com, 2004–2008). This technique made it possible to start and stop the whole system through a web interface. For the pen based set-up a server handed over the data from the sensor system (Polhemus Isotrak or SmartTech SmartBoard) to the client systems through UDP packages in order to enhance performance.

3.4. VIDEO RECORDING

To capture the actions of the users, all sessions were recorded on Mini DV tape with a Sony Handycam PC9. The videos were converted to standardised MP4 containers with MPEG-4 part 10 (that is, Advanced Video Encoding or H.264) video and AAC audio encoding of medium quality (22 kHz mono). A small application was developed to code the videos according to qualitative criteria. The data analysis was conducted with the 'R' statistic programming language (2006–2008, The R Project for Statistical Computing).

3.5. EVALUATION

This section elaborates the evaluation methodologies that were applied in this experiment. Firstly, the questionnaire and its relevance for a usability test of the AR setting are discussed. Secondly, it reports the methodology applied to quantify the data of the video account. The analysis of the data was conducted by using EDA test (Experimental Data Analysis). This methodology was chosen over a confirmatory analysis, as there has been no prior knowledge for evaluating AR systems like the one described. Therefore it would be arbitrary to elaborate discrete hypotheses for each problem and have them proven or rejected.

3.6. QUESTIONNAIRE

The questionnaire used was based on the questionnaire originally developed by Gerhard *et al.* (2001) to measure presence with an avatar aided VE. It was chosen as the most appropriate questionnaire for this experiment as it included aspects of direct collaboration, the presence of another person, involvement in the task and satisfaction with the system. Original items related to avatars were replaced by questions about the other participants. Other minor adjustments were made in order to address AR specific issues and to reflect the architectural task in the brief.

The choice of a presence questionnaire came from a large body of work that relates the term presence to factors of affordance (Conole and Dyke, 2004). It can be argued that an object providing an appropriate affordance, perceived or actual, contributes to the sense of presence. The questionnaire by Gerhard *et al.* (2001) originates from the 'Presence Questionnaire' (PQ) by Witmer and Singer (1998) which measures items like sensorial factors, distraction and realism. These were relevant in this experiment, so were kept in the questionnaire. Presence questionnaires are a viable source of measuring the involvement with a simulation system.

Gerhard *et al.* (2001) argue that awareness and involvement are essential for a sense of presence. Therefore, the more the system is removed from direct perception; the better a user can feel 'within' and subsequently work within that environment. For the experiment, it was necessary to look at the level to which the users obtained a feeling of working together and physically sharing the same environment. 'Spatial presence' is important as it could be fostered to ease the sharing of large datasets in collocated settings. A potential extension would be that an adapted technique could also involve distant parties in the AR environment transparently.

The final questionnaire measured 21 items for satisfaction (SAT), interface (INT), communication (COM), presence (PRE), awareness (AWA) and involvement (INV) (Table 1). A Likert scale from Agree (1) to Disagree (5) was used. Two additional items in the questionnaire gave the users the opportunity to judge the proposals they observed within the AR environment. Furthermore, there were four qualitative items, which captured opinions and problems that the users encountered with the system.

Table 1. Items of the questionnaire.

Item	Description	Count (reverse)
SAT	Satisfaction with the system	3 (0)
INT	Impact of Interface	2 (0)
COM	Communication between the users	3 (0)
PRE	Presence, feeling within	6 (1)
AWA	Awareness and distractions	5 (1)
INV	Involvement with the task	2 (0)

3.7. SUMMARY

Firstly, the application area determines the design and development of an AR system but secondly it also serves to reveal valuable insight for creating a generic software framework. The implementation used here is the result of several years of improvement and a focus on essential components for digital design. Beyond that AR is potentially an omnipresent technology in CAAD. To achieve this, the underlying concepts and technologies in CAAD need to change to facilitate the needs of AR. Collaborative design within CAAD systems and AR visualisations needs to progress beyond the exchange of files.

To make AR feasible as an extension of the digital design process, aspects like visualisation with NPR (Non-Photorealistic Rendering) are needed for schematic design representation. AR aided design tools need to connect visual appearance with a priori knowledge. This will enable designers to use AR tools ubiquitously within the design process. Syntactical means like low fidelity of appearance can convey a straightforward semantic utility.

Another aspect is the support of collaborative techniques. Currently, the technology still relies on rudimentary bandwidth for communication channels. Collaborative AR demands low-latency high-bandwidth networking, which ubiquitously embeds into the workflow. Low latency is needed, as geometry transformations of an AR environment are uncertain and unpredictable and therefore have to be updated continuously. High bandwidth is required to facilitate the distribution of large datasets common in architectural design and to use a composite video for AR rendering.

Overall, AR as a new medium for CAAD faces a technological challenge to merge the technology back into the design process rather than being an external entity. At the moment, the design process has adapted to CAAD. The opportunity of AR is to make CAAD in-situ and ubiquitous. The conventional centralised and data driven approach inherited from the early days of digital engineering needs to be migrated over to a transparently distributed on-demand system which supports real-time visualisation. This study demonstrates an approach to utilise AR technology in a design setting and investigates human factors in regard to the design process.

4. Results

This section reports the raw data collected. The statistical analysis was conducted with a standard confidence interval at the 95% level.

4.1. SUBJECT SAMPLE

Overall 28 students and staff from The University of Hong Kong participated in the experiment. Ages ranged between 22 and 40 years. The sample consisted of 9 female and 19 male participants, which were evenly distributed within the two groups. The majority was of Chinese origin. Seven were of Caucasian origin. These measurements were not part of the questionnaire but were recorded in personal notes for each trial.

None of the users had encountered the system before. Four users identified themselves as practitioners, the rest as students. Overall 27 users had experience with design tools, with more than five years of experience. Six users had previously used Immersive Technology, three had more than five years experience. Overall, 14 users used the pen setting, five female and nine male. The cube setting had the same number of users, with four female and ten male.

4.2. TIME MEASUREMENTS

The users were asked to finish both phases of the brief in 20 min. The users could stop earlier if they felt that they fulfilled the brief. Predictably, the time for completion does not vary between the pen and the cube conditions. The users spent the same amount of time in both conditions, with a mean time of 16 min. The standard deviation of the overall time taken is high. But with and without outliers the difference was insignificant. The pen condition had a standard deviation of sigma = 4.1 min (with one outlier removed sigma = 1.8 min) and the cube condition a standard deviation of sigma = 3.9 min (one outlier removed sigma = 2.4 min). Removing the two outliers halved the variance. The two extreme values were a maximum of 23 min 30 s and a minimum of 11 min 28 s.

The design creation tool, pen or cubes, forced the users to follow different work timing. The difference between the two is statistically significant (t-value: 2.27, p-value: 0.04).

4.3. PERCEIVED PERFORMANCE OF INTERFACES

Two items in the questionnaire assessed the impact of the tool on the perceived performance of participants. The users judged the overall performance of the tool in regard to finalising the task. The data show that these items were quite reliably reported. The observation interface was the same for both groups and the data confirmed that the users unanimously agreed upon the perceived performance.

For the creation phase the data reported a significant difference between the conditions. The cube interface performed slightly better with statistical significant difference (t-value: -1.712, p-value: 0.01). The data show that there is a relatively large standard deviation of sigma = 1.53 in the cube condition.

4.4. SATISFACTION

Unlike perceived performance, satisfaction measured whether or not the users perceived the system as satisfactory for their needs. None of the items about satisfaction corresponded between the two conditions. Also in regard to the

HMD the data were consistent and therefore confirmed the similarity between the conditions.

4.5. PRESENCE

Presence is a key measurement factor for simulation environments. It measures the perceived level of immersion of the users within a system. Presence goes beyond the concept of immersion, which is purely technical. There are various categories of presence. Spatial presence describes the 'being with' of a user in regard of the space of the simulated environment. Social presence reports in relation to a user 'being with' avatars, synthetic entities or real persons. Object presence is the state of 'being with' an object in the same space and shared with others (Witmer and Singer, 1998).

Presence is important for this experiment as it is closely coupled with the concept of affordance. The more an interface, or a whole set-up, can be used instantaneously, the more likely it is that the user will not be disturbed in getting 'into' the system. The questionnaire reflects this granularity in regard to concepts of presence with items targeting different sub-categories of presence (Table 2).

Table 2. Presence items in detail with categories of presence.

Questionnaire item	Category
PRE1 (The feeling of presence inside the simulation was compelling.)	Spatial
PRE2 (The other participant was naturally within my world.)	Social
PRE3 (I was so involved into communication and the assigned task that I lost track of time.)	Social
PRE4 (Events occurring outside of the simulation distracted me.)	Spatial
PRE5 (My senses were completely engaged during the experience.)	Spatial
PRE6 (The other participant pointed at objects.)	Object

The combined items for presence did not yield any findings. The two outliers are the two users who reported Simulator Sickness AR system. One item which explicitly addressed presence as it pertains to virtual objects within the same environment showed a statistically significant difference toward the cubes (t-value: -2.33, p-value: 0.028).

4.6. AWARENESS

Awareness measures the degree of immersion to which the user is exposed. This factor influences presence but can also be seen as a separate more technical factor. Awareness in the questionnaire was, similarly to presence, measured on two levels, social and object awareness. None of the items produced a difference in response. The first item showed that users could still easily differentiate between real and virtual objects and that they were aware of the actions of the other user.

4.7. INVOLVEMENT

The involvement for both groups was high and consistent across the conditions. The two items in the questionnaire targeted different levels of involvement. Whereas one item measured the relative engagement, the other measured an absolute value. No statistically significant difference for gender could be found in the data between both conditions.

4.8. PERCEIVED COMMUNICATION

The items in this category gauge the users' perceived impedance of communication. In total, the communication was perceived by the users as unhindered (Mean: 2.4, sigma = 0.97). A difference has been observed on an item that directly asks for the back channel communication. The cube condition performed slightly better than the pen condition. This difference is close to being a statistical tendency ($t = -1.41$, $p = 0.17$).

4.9. COMMUNICATION PATTERNS

With the above measurements, only the perceived factors of communication can be assessed. Post-rationalisation is an apparent problem in assessing aspects of communication.

Patterns of communication can be analysed through post-experimental coding methods like the one proposed by Kraut *et al.* (2002). This specific coding is used for analysing videos for instances where users refer to objects, name their position or agree. These three items have been divided into sub-items specifying the context of the utterances. Thus, one can capture the frequency of references to an object directly or through the spatial context. Furthermore, utterances referring to the position or the position context, agreement on general ideas or agreements on actions (i.e. behaviours) are coded. There was no significant difference in the overall amount of utterances between the pen group (545 or 2.4 utterances/min per person) and the cube group (521 or 2.3 utterances/min per person). Nevertheless the communication showed different properties through the frequencies of utterances. Histograms were used to illustrate the difference visually (Figure 3).

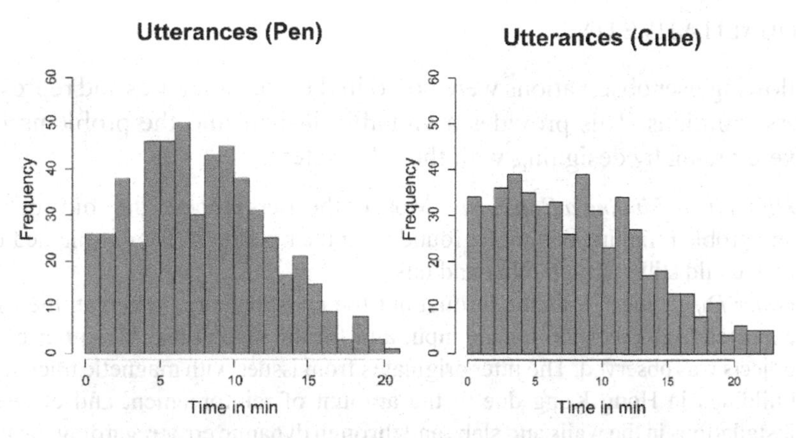

Figure 3. Utterances compared in histograms.

To investigate the communication patterns in more detail the sub-items were measured regarding their density (Figure 4) and density diagrams over time were used. (Density diagrams graphically represent the bandwidth used for a certain point along a time-line. The y-axis is the percentile of the complete conversation 'spend' at time for utterances. The x-axis is the time-line of the experiment.)

The absolute references showed a distinctive pattern for both settings with two peaks and a trough in between. The first peek accounts for the groups deciding on one of the proposals. The trough denotes the users changing the interface. And the second peak the discussion about the extension. Therefore the pen users used considerably more bandwidth in their conversation to discuss the design proposal. The other graphs show that for the other sub-items the conversation patterns are almost identical.

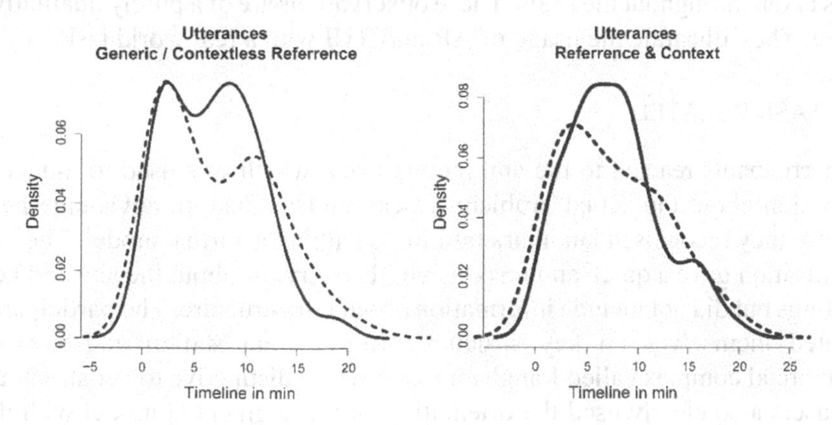

Figure 4. Differences in utterances. The solid line denotes the pen condition, the dashed line the cube condition.

4.10. QUALITATIVE DATA

The following user observations were noted in the questionnaires and represent the users' opinions. This provides a valuable insight into the problems that users were facing by designing with the AR system.

- *Difficulty in Finding a Consensus*: Most of the users reported they did not have any problem finding common ground with their partners. They explained that they could still see each other and talk.
- *Other Difficulties*: A theme throughout the questionnaires was that, the users expected high precision for the input and for the modelling. Also jitter of the objects was observed. The jitter originates from issues with magnetic tracking in buildings in Hong Kong due to the amount of reinforcement and electrical installations in the walls and slabs and through dynamic error regarding the pose recognition in AR Toolkit.
- *Significance of Experiencing AR*: Users referred to the directness of the cube design tool as a positive aspect but comment that it is too rudimentary to use for detailed design work. Some noted here the lack of visual vividness of the AR models in comparison to the surrounding real environment. Some users asked for a higher degree of realism with textures and material.

Throughout all of the trials in one way or another, poor picture quality was mentioned as a hindrance if the system was to be used seriously. This was not a surprise as it has been mentioned earlier by Campbell and Wells (1994) and the HMD technology has not improved significantly through the years.

5. Observations

The following observations are an account of a review of the video account and notes taken throughout the trials. These observations are of a purely qualitative nature. They illustrate the usage of AR and TUI with a real world task.

5.1. TASK RELATED

All participants reacted to the ambiguous brief, which was used to induce a discussion about the actual problem. Local students had an advantage here because they recognised landmark buildings within the virtual model. The AR visualisation gave a quick and responsive 3D overview about the site and key buildings but did not include information about infrastructure. The participants oriented themselves on key structures like a train station and a large commercial complex called Langham Place with a distinctive tower structure. The users also clearly used the orientation of the augmented model with the north direction facing away from them.

From the discussion protocol one can extract that the users were quick to decide that their proposals were a counter-balance to the size of the existing development on Langham Place. This was observed four times in the 14 groups. Other groups also were aware of the size and moved the proposal into the existing structures and explicitly stated that they wanted to tear down other parts in order to integrate it into the urban fabric. The spatial arrangement and the visual cues of the AR environment clearly helped to orient the users and provide a rough estimate for urban massing. All this intrinsic information for the first design steps was extracted without consulting plans or other media.

5.2. DESIGN AND SPACE

All groups, without exception, decided to move the design proposals away from the initial site. One group challenged the system by stacking cubes underneath the proposal to make it higher and exposed to the surrounding area. The whole digital model was un-textured, with light grey material with a matte shader. These rendering settings were chosen to resemble the appearance of foam-board and cardboard models as closely as possible without imitating them. Although, the users did realise it was an initial design analysis task, several wished to have more visual richness.

One group encountered difficulties in finding common points with higher precision on the table, which was caused by optical distortions common in video-mediated AR systems. However, they found a crutch to overcome the problem in a green adapter, which was accidentally placed near the table and not part of the set-up. Utilising this artefact as a coloured marker helped them to orient precisely and discuss layouts and landmarks. This unintentional tweak is interesting as it could be helpful in other AR applications as a tangible and spatial bookmark.

Most users of the pen setting went beyond the brief and created more cubature than needed. Similarly the users of the cubes used all of the cubes to work in parallel and in different areas. This behaviour clearly indicated the exploratory nature of the AR technology and also questions the heavily restricted brief that had been given to them. But also signifies the low barrier that users encounter in relation to exploring spatially with AR technology.

5.3. ORIENTATION AND USAGE

There was no additional information given about the experiment. The users relied solely on the knowledge they retrieved while working with the experiment set-up. Some users asked for the north orientation as they were used to this on drawn plans. The north orientation was important for them to orient the proposals and the extension. This provided interesting clues for future implementation. The set-up followed the north-up convention found in plans.

In general, only a few users were concerned about the feasibility of a purely formal investigation in an urban design setting. They tended to extend the discussion to include the use of the buildings but in general viewed the system as a brainstorming aid. Local students especially, had elaborate ideas about the usage and impact of the buildings retrieved through their knowledge of the place. The AR setting helped to give spatial cues about the place without overloading participants with contextual information. A formal investigation is most unlikely to generate useful architectural proposals. However, most users reported that they found it to be a valuable brainstorming tool.

6. Discussion

The objective of the experiment was to reveal differences between TUI and a conventional VE by comparing a pen-shape interface derived from VR applications with a typical AR interface. Context for the experiment was an architectural design setting. The experiment provided insight into an AR aided design process, the application of the interfaces provided and the effects on communication.

6.1. CONSTRAINTS AND LIMITATIONS

The above experimental setup introduces some constraints that are important to consider in regard to the transferability of the experimental results. First, the targeted group of users consisted of students and practitioners in architectural design with a priori knowledge of digital media. The sampling of the subjects produced dependent pairs.

Two users were paired for the test and it was necessary to ensure that the participants were willing to collaborate and communicate. There is no indication here that the same AR application setting could scale for more users or be usable for a remote collaboration. There was no control applied on the pairings. Furthermore, a convenience sample was used, only postgraduate students and practitioners available at the Department of Architecture at The University of Hong Kong participated in the experiment. The test was voluntary and the participants did not receive a monetary compensation or other incentives. Most of the postgraduate students had experience in overseas practices and universities, as this is part of their curricula. They were confident in communicating in English.

All of the users had experience of digital design tools and therefore easily picked up the design task. This was beneficial for the experiment and it is likely a scenario expected in the future for most architectural designers. It also highlights an important opportunity to provide new tools for the design process like AR. The data report a high percentage of users were working with VEs.

This was expected, as designing for and inside of VEs is part of the curriculum for architecture students at the Department of Architecture at The University of Hong Kong. Therefore, the sample represents a new generation of architectural designers exposed to and savvy with immersive technology. Fostering this potential is in the context of the experiment an opportunity, as the results are not overshadowed by the novelty.

6.2. DESIGN ACTIVITY

Based on the data presented in the previous section, users in both conditions took the same time to finish the task. In both conditions the participants finished earlier, however few groups took significantly longer to complete.

The time frame for the design phase for the trials differed significantly. The cube users were using about double the time for design than the pen users. From the videos one can observe that the users in the cube condition spent more time on quickly discussing different approaches, whereas the pen users were cautiously placing and moving cubes in the site. This difference can partly be accounted to the different affordances of the tools. The users with the pen needed to change to a different device. This change resulted in an adaptation to a different affordance of the tool in order to create objects. For the cube condition, the users could perform the creation phase of the brief with the same learned knowledge from the first phase. Furthermore, the cubes as spatially multiplexed input devices were easy to share. Users with the pen needed to take turns with the input device. Only one out of the seven pen groups took turns with the pen, the others decided on a 'pen operator.'

For collaboration, this has an interesting effect. The users of the pen condition were relying on the communication and also showed that there were considerably more exchanges (or utterances) regarding the references, which clarified the site for the extension. This observation therefore confirms an observation made by Vera *et al.* (1998) who found that low bandwidth tools can enhance the communication. The current work extends this notion into the domain of spatial manipulation interfaces.

From an observation of the video it becomes clear that the pen users spent more time communicating about details. Additionally, pen users 'held off' the creation phase of the task (that is, extension building) to verbally discuss the actual problem in more detail.

The cube users discussed aspects of the design almost in parallel. That means they went for a 'trial-and-error' strategy. This reduced the amount of spatial related communication exchanges but supported a faster iterative exploration. Throughout the experiment, cube users were considering a multitude of design decisions in breadth and in shorter time.

6.3. COMMUNICATION AND INTERFACES

The questionnaire reported no differences for the combined items of the perceived communication. A slight difference (statistically insignificant) did show for the back channel. The users in the pen interface apparently had less response from their partner. This hints towards a difference in regard of the interface multiplexing Fitzmaurice and Buxton (1997). The cube interface is space-multiplexed; therefore each user can independently choose a tool and worked with it. The video account clearly sustains this, as almost all users in the cube condition chose to work in parallel. The pen condition however is time-multiplexed; each user has to wait for his/her turn. The participants mostly left the pen with one user. Subsequently, users had to put more effort into communication and perceived a lack of a back channel.

The most important element for understanding the communication is the video account. The different levels of communication are visible in the histograms of the communication items. Both conditions changed to another tool; however the pen interface used another interaction technique. It also shows a distinctive pattern for the agreement on behaviour. The pen condition relied heavily on one users talking to the ‘pen operator’ about their intentions and confirming actions in the virtual scene. Cube users worked independently and therefore could leave out constant agreements. Therefore the structure of the user communication was affected by the paradigm shifts for the user interface and by the interface change per se. Consequently, the effects on communication can not be compared directly with the perceived performance because the actual design strategies differed.

The different shapes of bandwidth usage between pen and cube condition serves as an indicator of the impact of a perceived affordance on communication (Figure 4). The second peak in the pen setting indicates that the users both had to re-negotiate and additionally put more effort into communication in order to interact with the system.

The cube condition allowed users to work simultaneously on the task but in the pen condition the users often ‘held back’ the design for the extension and discussed their strategy in advance.

For support of architectural design this has two effects; the low-bandwidth single-user approach does support a higher communication exchange, but the high-bandwidth parallel support ends up in a higher frequency of design attempts and longer time spent on the design.

6.4. INTERFACE PERFORMANCE

The data for this factor confirms the reliability of that which had already been noted by other research (Nash *et al.*, 2000). Both groups used the same interface for observing the proposals and the data reflects their choice unanimously.

Overall the data shows that the participants found the interface for creating the proposals less suitable than the observation interface.

For creating the extension (i.e. the creation interface) on average the users preferred the cube interface as being more suitable to finish their task. There is also a difference in the variance between the conditions regarding the interfaces. The cube condition showed a standard deviation that was between 1.5–2 times larger. This variance originates from users in the cube condition spending a significantly longer time on the design task. This meant that they were therefore more likely to encounter situations where the occlusion was becoming a problem. Therefore, it can be assumed that TUIs can be beneficial for exploring variants of a design as long as the visual fidelity is given. But it does not necessarily mean that non-TUIs are more efficient in terms of faster completion or more precise interaction.

The pen interface, due to its single-user attributes forced users to discuss the design to work in a team. It confirmed that a preferred interface is not automatically the most efficient one in terms of design communication. For the early phases of design it can be beneficial to explore more variants, but it does not encourage more communication, which is crucial in those situations. It is yet unclear if the overlay of affordance of the pen is a hindrance or just the paradigm shift of affordance. Users in this experiment did not relate the use of the pen to a cube generation mechanism even though they were briefed and tested the cube creation technique in the warm-up phase.

6.5. SATISFACTION

The users, as shown earlier, were mostly satisfied with the system support for the task. There is an insignificant advantage for the cube condition which can be accounted for by the easier working of the interface. The score of the overall system also showed that users could distinguish between the user interface and the installation. A factor that was expected to show in the results was the weight of the HMD. Taking into account that one of the HMDs had serious comfort issues it was surprising that it did not effect the satisfaction, nor could any correlation be found between the HMD and the questionnaire items. However, users reported discomfort in the qualitative section. They clearly stated that systems of this kind will be unusable for them unless the displays change. This also reinstates the observation that users tend to endure great discomfort as long as they are given a novel and interesting task.

6.6. SPATIAL, SOCIAL AND OBJECT PRESENCE

The experiment did not show any significant difference for spatial and social presence. This is expected as both conditions use an identical AR environment with only the input user interfaces differing. Therefore, the overall system design was comparable in that respect.

A difference however was observed in relation to object presence in which users in the cube condition reported a statistically significant higher perceived presence in regard to virtual objects within their environment. This confirms the expectation that users with TUIs were advantaged because of the physical and virtual entity match, visually and through touch. Additionally, the pen condition used a Polhemus Stylus which overlays the perceived affordance of a pen with a button to press in order to create an object, which was not obvious to most users.

It can therefore be assumed that TUI which replicate very basic functionality can afford a more direct relationship to the virtual environment. Hence, they can yield a higher degree in perceived presence. Transferring functionality from the two-dimensional domain to three dimensions (i.e. the Polhemus Stylus) will make the device more adoptable but only if it stays in the same interaction realm.

The experiment also confirmed earlier research (Regenbrecht and Schubert, 2002), which demonstrated that low-fi media like books can actually produce higher presence as long as their interaction paradigm matches the task at hand. The addition of an extension through virtual bricks seems to be more obvious than through a point-and-click interface.

6.7. AWARENESS

Contrary to what users reported for the presence items, the data for awareness showed that the users were clearly aware of being within a simulation where virtual and real objects coexist. Thus it was not necessary for the users to feel completely immersed in order to perceive presence. This complements the findings of the presence scores and confirms other research that reports that low immersive technology can create high presence. Interestingly, users felt that they were pointing to the same objects, even though they were aware that they were interacting with a virtual environment where objects do not really exist as one entity.

The experiment looked at explicit interaction techniques only. In future it would be interesting to look at context aware interaction techniques. Context aware interaction techniques, like gaze-based interaction can potentially provide another perspective as the simulated environment directly responds to unconscious actions of the user. Potentially there is a correlation between awareness-based interaction techniques and awareness in relation to presence.

6.8. INVOLVEMENT

The experiment showed that users were unhindered by technological shortcomings in their ability to get involved with their peers. Overall, users scored as having high levels of involvement. Future research needs to investigate whether or not this involvement can be influenced by other factors. High involvement also underlines the necessity of using a real world problem for assessing the impact of a design interface. Users reported in the questionnaire that they see AR as a valuable tool for becoming more involved with design work and for discussing ideas with peers, although they acknowledge the infancy of the technology.

6.9. RELEVANCE

The experiment used an abstracted design task in order to measure usability in a praxis relevant setting. It is therefore transferable to other domains in terms of the data collected for AR and TUI aspects because it did not involve the users in an artificial experimental set-up and kept influences of architectural design methodologies to a minimum.

The software had limited design capabilities in order to keep both settings comparable. The intent, which was reflected in the brief, was a formal design investigation in the early phases of a project. It therefore did not use elaborate visualisation or simulation techniques nor did it provide any means of storing a working status even though that is technically possible. In sum, it was a brainstorming tool for designers, which users enjoyed and was recognised as such.

Some of the technological aspects were constrained by the venue and resources available at the Department of Architecture at The University of Hong Kong at this time. More sophisticated methods of spatial registration are available for AR and could enhance the experience with the system. The participants were all affiliated with The University of Hong Kong and there were no control measures applied about the pairing of the users.

The experiment also confirmed an observation made by Dennis *et al.* (1999) about gender differences being non-existent in relation to involvement in the system. However, this finding needs to be re-confirmed with a similar AR setting controlling gender distribution explicitly.

7. Summary

The main findings of this study are

- Cube users (TUI) took a longer time designing to explore more ideas.
- Pen users (Direct VR) took less time but for fewer design iterations.
- The perceived performance of the interface was better with the cube condition.

- Cube users felt more present with objects in comparison to pen users.
- Communication was unhindered except for small difference in the back channel in the pen condition.
- The communication pattern differed for the two conditions, specifically concerning referencing objects and agreement with the peers within the simulated environment.
- Users were satisfied with the system even though they complained about the HMDs.
- Participants were aware that they are in a simulated environment. This did not affect the perceived presence.

This experiment showed that tangible user interfaces can have an impact on presence, performance and communication. It can be concluded that the usability of TUI in AR is suspended between the tangible component, its relation to the actual task and the software implementation component which creates the connection between real and virtual object. Balancing these components is a fragile process, but is the key to usable AR-based design interfaces.

Pen-like devices have been valuable for VEs. As Piekarski and Thomas (2002) pointed out, VE and AR share common ground in relation to HCI issues for input devices. However, more aspects, like the communication patterns observed in this study attest to the necessity to revisit this assumption. There are limitations of using an abstract concept like the pen and also the TUI. Although, the pen itself is an omnipresent tool in architectural design, this experiment showed that other interaction paradigms can be valuable for certain tasks. Whereas the pen condition let users work in detail with detailed consideration, the cube TUI helped users to explore a broad variety of solutions in parallel. This experiment contributed new knowledge about interfaces and their impact on the design process. It opens new questions about the adoption of new interfaces for the design process. Is there a learning effect over time and does the interface affect the actual outcome?

AR aided design needs a novel way to provide both constrained and freehand spatial input. The early stages of drafting and brainstorming require an unambiguous and conceptually reduced interface. The experiment showed that the interface can support or prevent certain design strategies, and affects the design process.

8. Conclusion and Future Research

Technologically, AR has a long way to progress. It is at this early stage that architectural design should embrace it as an opportunity. For architectural design, the potential of AR goes beyond a visualisation tool. AR is a communication medium. The combination of real and virtual elements makes an ideal context for a design team to examine spatial problems in a collaborative setting. Boundaries between real and virtual will be pushed further than we can predict today but it is necessary to remind the designer that AR and the supporting infrastructure are a utility and not a push-button solution for creative output. Creative processes need to be in the control of designers aiding them to develop their creativity further. For this to occur, emerging technologies like AR need to provide interfaces with a maximum of freedom and a minimum of pre-programmed logic in order to allow the designer to maintain rather than constrict creativity during the design process.

Already, the architectural profession has embraced digital design as one of the necessities of the university curriculum. The challenge is to give the media savvy designer the right tools at hand because at present, digital design tools do not usually support co-located collaborative design processes.

For future research, several questions and suggestions arise from this study, in particular to assess whether or not the design outcome is comparable when different devices are applied in an AR based setting. This was out of the scope of this experiment design and further investigation of this factor will require a new and more refined experimental setting. The experiment demonstrated that the tasks and their contexts are relevant even if they are abstract.

Surprising factors can influence design decisions. For example in the experiment local knowledge was an influencing factor. One of the findings confirmed that users could relate the rather abstract urban design task with the real problem. They related their earlier acquired knowledge of the design studio with the AR environment. This is a positive finding but raises the question of how this knowledge can be made more accessible to peers if the users are working in a remote collaborative setting. Another important investigation is that of fidelity. The introduced software prototype already supports physical interaction between virtual and real content. It needs to be ascertained if the pure virtual simulation of physical behaviour in an AR design environment will have an effect on the design outcome and if it will have an impact on other factors of collaborative working.

Design interfaces, aided by AR using direct interaction can provide a valuable asset. However, the balance of abstraction and functionality is a subtle issue that will vary from task to task. The use of design systems is currently overshadowed by the simplicity of traditional tools like pen and clay. The challenge is to foster the advantage of physical engagement for digital design, enabling a new kind of shared immediate interactive creativity.

Acknowledgements

The author wishes to thank Thomas Kvan, Marc Aurel Schnabel, Mark Billinghurst and Justyna Karakiewicz for their invaluable contributions, critique, encouragement and patience. This research was made possible through a PGR grant of the University of Hong Kong.

A TECHNOLOGICAL REVIEW TO DEVELOP AN AR-BASED DESIGN SUPPORTING SYSTEM

JIN WON CHOI
Yonsei University, Korea

Abstract. The construction industry consists of broad areas like architecture, urban structures, landscapes, bridges, harbours, roads, water supply and drainage facilities, airports, et cetera. However, these need to be classified for the application of AR technology. To address this problem, we first suggest a new classification according to building types, with the sub-areas of construction industry, main participants, issues and activities at each design and construction stage. By integrating the core technology of Augmented Reality, we can suggest some application systems that can be applied to various areas of construction industry based on the development of a platform supporting design. A design support system using realistic AR techniques can bring about a change in the IT industry directly, and should indirectly have a significant impact on the traditional construction process.

Keywords. Mixed Reality, Augmented Reality, Construction Management System, Construction Process Simulation, Life Cycle of Construction Industry.

1. Introduction

Since the 1960s, when wireframe was originally developed as an early form of computer graphics, computer technology has come a long way, yet it still needs to develop more practical approaches to the representation of augmented and mixed reality.

Several recently published studies on the subject of augmented (or mixed) reality have stated the philosophical and aesthetic importance for a variety of fields, but especially for architecture, which takes a serious view of visual output. Despite various attempts, the application of augmented reality to the area of architecture remains at the conceptual level unlike practical solutions developed in other areas such as the current CAD system.

X. Wang and M.A. Schnabel (eds.), Mixed Reality in Architecture, Design and Construction, 53–74.

Although related applications and component systems of augmented reality are becoming more prevalent, we still do not have a clear understanding of how this technology will function in the construction industry and subordinate areas. Given the basic characteristics of the construction industry, it is not surprising that there are problematic effects associated with the complicated process of construction and its relationship to stakeholders in civil engineering, urban design, and interior design. In order to develop technologies effectively, based on these characteristics, it is necessary to offer some solutions corresponding to the specific needs of each subordinate industry. Hence, the necessity for a construction management system based on augmented reality, which has the capability of offering interaction and presenting richly detailed information using computer graphic technology. Further, it is anticipated that such an augmented reality system will facilitate the growth of the construction industry, as has been the case with other technologies in the fields of medical care, aviation, and education. It is critical therefore, to perform fundamental studies on the use of augmented reality in potential construction management systems. To begin the process, this article presents research related to the potential application of augmented reality in the construction industry.

2. History and Application Domains

2.1. SOCIOLOGICAL & ECONOMIC EFFECTS OF AUGMENTED REALITY

2.1.1. Description and Perspective of Augmented Reality

Augmented reality (or mixed reality: AR or MR) is an advanced information technology that links digital information such as computer graphics, sounds, haptic systems, scents, et cetera and real objects in physical environments in real-time, which improves user thereby improving user interactions (Azuma, 1997; Azuma *et al.*, 2001).

Currently, augmented reality systems are considered important technologies capable of offering multimedia content more effectively and accurately than conventional techniques by making information in physical environments simpler.

Ivan Sutherland first introduced augmented reality, when he developed a see-through HMD (Head Mounted Display) in 1960 (Sutherland, 1968) and the realism of this mixed reality (MR) depends on a range of parameters. As Milgram and Kishino argue, the extent of realism relies on the controlled use of the virtual-reality continuum, the degree of modelling, and the viewpoint and hardware capacity (Milgram and Kishino, 1994). More recently, Gartner Research (2006) classified mixed reality technology as a technology trigger, and further described it as an influential technology (Figure 1).

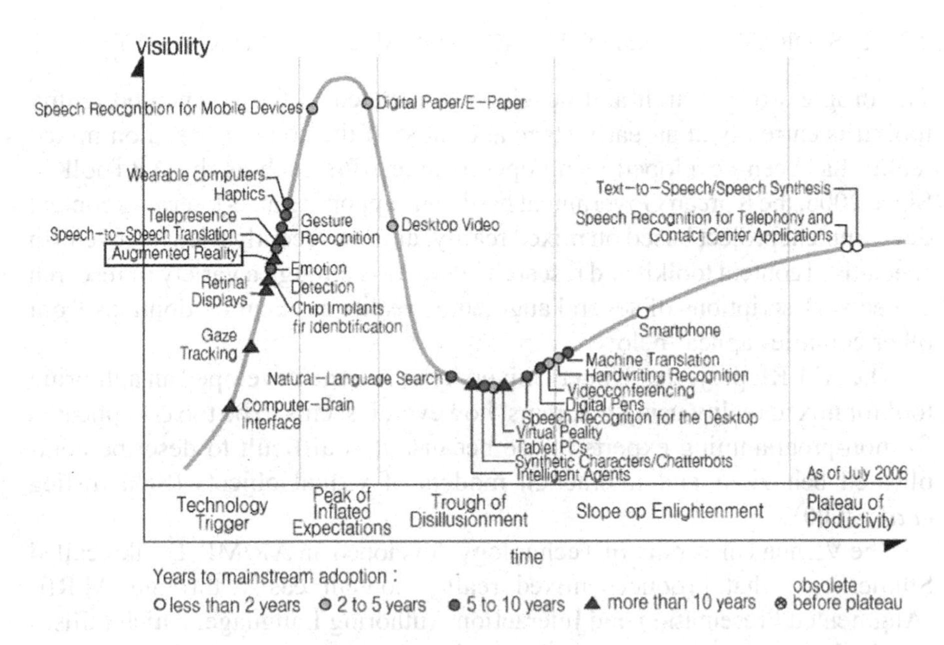

Figure 1. Hype cycle for emerging technologies (Gartner Research, 2006).

2.1.2. Design Support Systems

Augmented reality systems have already been used in design both technologically and theoretically. Several examples present some of the possibilities of augmented reality: a tangible interactive platform that consists of 'STC tools' – a sketch facility based on an interactive concept exploration software application, and; ARdesk – a video projection-based augmented reality desk to assist designers (Nam, 2005); a Spatially Augmented Reality Design Environment (SARDE) that can be used as a design visualisation tool in the field of interior design (Chen and Chang, 2006); tangible interaction design for a cooperative urban design system (Lee, 2003); and a Workbench study for urban design using augmented reality systems (Ishii *et al.*, 2002).

2.1.3. Cooperative Design Systems

In the domain of cooperative design, there are several good research examples such as *Computer Representation to Support Conceptual Structural Design within a Building Architectural Context* (Mora *et al.*, 2006), which suggests using a computer representation called StAr to support cooperation between architects and engineers for structural design. Additionally, *Real-Time 3D Human Capture System for Mixed-Reality Art and Entertainment* (Nguyen *et al.*, 2005) enabled people to have a teleconference and work together using augmented reality.

2.2. HISTORICAL TRANSFORMATIONS OF AUGMENTED REALITY

The progress of research and development related to the augmented reality toolkit is currently at an early stage and most of the content based on mixed reality has been developed using open source APIs such as the ARToolKit. Since 2006, the Korean Government has been supporting an e-Learning content development project based on mixed reality, and has recently tried to develop a specialised content toolkit and research interaction through a variety of research projects. Descriptions of several augmented reality application domains from other countries appear below.

The AMIRE project at Upper University in Austria developed an authoring tool for mixed reality using diagrams; however, it seems to be too complicated for non-programming experts to use because it is difficult to describe complicated behaviour and interaction models of virtual objects (Schmalstieg *et al.*, 2000).

The Vienna University of Technology developed an AR/MR Toolkit called StudierStube that produces mixed reality content easily through APRIL (Augmented Presentation and Interaction Authoring Language, which utilises standard content expressions and scripting. However, this has been limited because it is based on a desktop/HMD, rather than on mobile appliances Figure 2 (Graz University of Technology, 2006; Kato *et al.*, 2008; Reitmayr and Schmalstieg, 2003; Reitmayr and Schmalstieg, 2004).

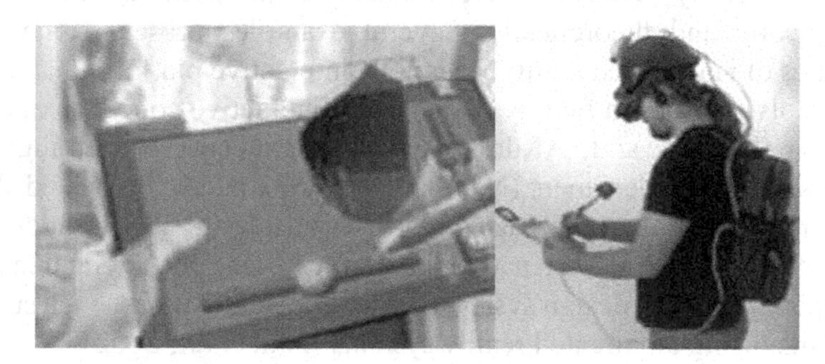

Figure 2. StudierStube structure and example of contents authoring (Graz University of Technology, 2006).

The HIT Lab in New Zealand developed OSGAR by combining an open-source toolkit called OpenSceneGraph (Osfield and Burns, 2006) as well as ARToolKit (Kato *et al.*, 2008) (Figure 3). Accordingly, several calculated resources such as special effects, shadows, collision detection et cetera for virtual reality can be used for mixed reality as well (Reitmayr *et al.*, 2005; Kato *et al.*, 2008). Nevertheless, it is still coordination at a toolkit level and hard for non-programmers to control.

Figure 3. Output of OpenSceneGraph and ARToolKit.

Figure 4 shows a structural drawing and a demonstration scene of an augmented map system developed by Cambridge University. This system is an augmented paper-based artefact that projects digital information onto paper maps using a special matching technology that allows it to take advantage of both paper based and digitally based maps (Lee *et al.*, 2005).

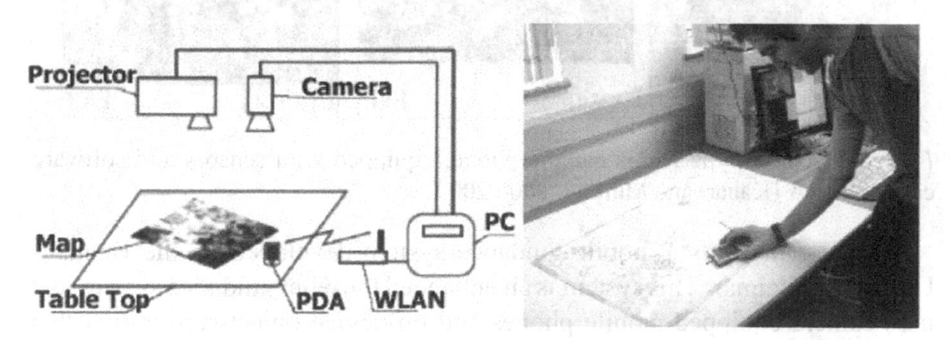

Figure 4. Augmented map system.

The Digital Experience Lab at Korea University developed an authoring tool for augmented reality and performed an experiment based on Occlusion (Figure 5). This system can be controlled three-dimensionally through generating, modifying, shifting and detecting position and motion behaviour et cetera (Lee *et al.*, 2005; King *et al.*, 2005).

The Mobile Augmented Reality Applications project at the Nokia Research Centre explores the possibility of utilising camera-equipped mobile devices as platforms for sensor-based video and see-through mobile augmented reality. Figure 6 shows a hand-held MR device and images of the See-Through Mode

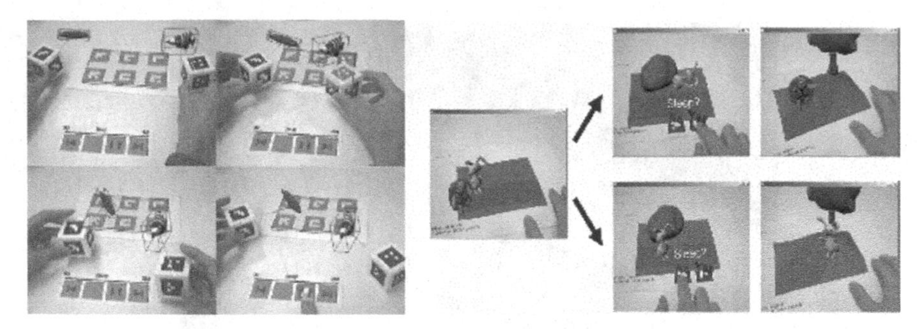

Figure 5. Authoring environments based on Occlusion.

and the Map-View Mode. This system uses accelerometers in all three axes to determine orientation, a tilt compensated compass for heading, and GPS for positioning (Park and Lee, 2004).

Figure 6. A prototype Nokia camera phone, equipped with sensors and software called MARA (Kähäri and Murphy, 2006/2007).

Figure 7 presents a mobile guidance system developed by the Bauhaus University Weimar. This system is an enhanced museum guidance system that uses camera-equipped mobile phones and on-device object recognition that provides additional location and object-aware multimedia content to museum visitors.

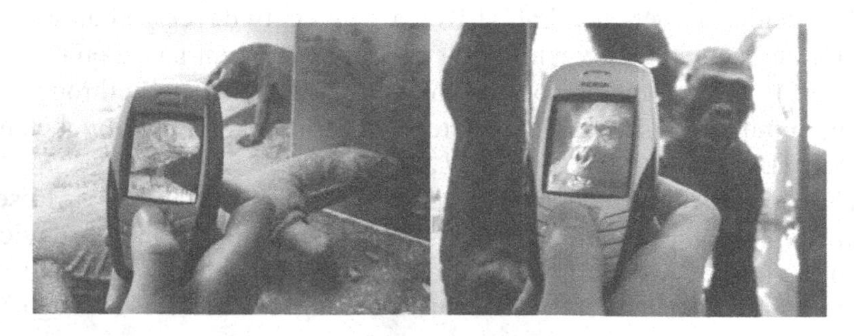

Figure 7. Application of PhoneGuide in a museum.

Additionally, described below are some augmented reality applications in the construction area.

The Archmedia Lab at Yonsei University researched a mobile augmented system called C-Navi. Researchers developed this system through a variety of processes. First, using 3D construction models, the construction schedule and the site database were established. Next, position and orientation tracking modules based on a GPS navigation system called 'Cityscape' were developed and the AR-based visualisation module was developed using a modified ARToolKit. Then the core engine of C-Navi was created through a process of integration; however, some calibration was needed to adjust for accuracy required to match two worlds. Finally, four extension modules composed of a 4D CAD module, a solar analysis module, a view analysis module, and a 3D utility module were generated and applied (Figure 8).

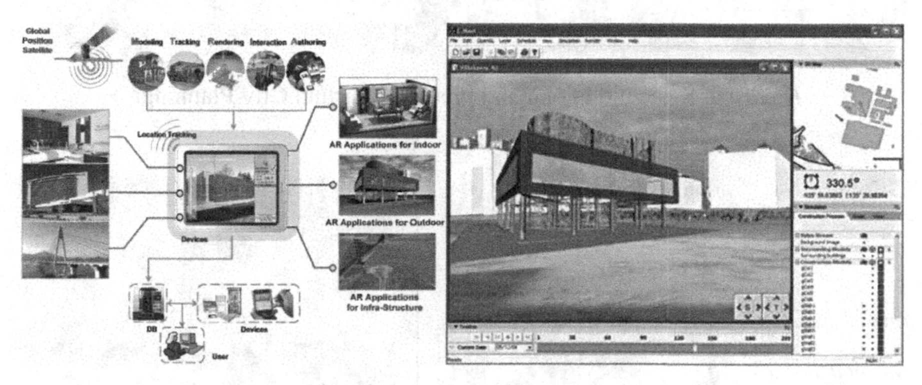

Figure 8. Schematic diagram and user interface of C-Navi.

The Department of Digital Media at Ewha Woman's University in Korea conducted a virtual city planning project that set out to develop a virtual reality system for urban design focused on immersion and interaction in virtual environments. Since urban design requires prior evaluation and quick feedback, the goal of the virtual city planning project is to produce an urban design simulation based on visualisation and data processing. Moreover, in order to render in a real-time manner, the virtual city planning system offers dispersion rendering environments and focuses on visualisation and LOD (Level of Detail) technology. Currently, Virtual City Planning consists of a virtual city manager with the capacity to develop designs based on GIS, and a tangible user interface manager that is able to detect makers, as well as offering schematic map data (Figure 9, 10).

Mobile Augmented Reality Systems for exploring the Urban Environments' is a project developed by Columbia University (Feiner *et al.*, 1997). The aim of the project is to conduct research into a prototype that enables users to find a path by obtaining information from special objects such as buildings or monuments to create a campus tour guide (Figure 11, 12). In addition, another project called Digital Design Mock-Up was developed as a MARS application. The system enables users to see mock-ups in real environments based on an augmented reality system, to check for contextual appropriateness such as for congruence with the surrounding buildings.

Figure 9. System diagram and image of Virtual City Planning.

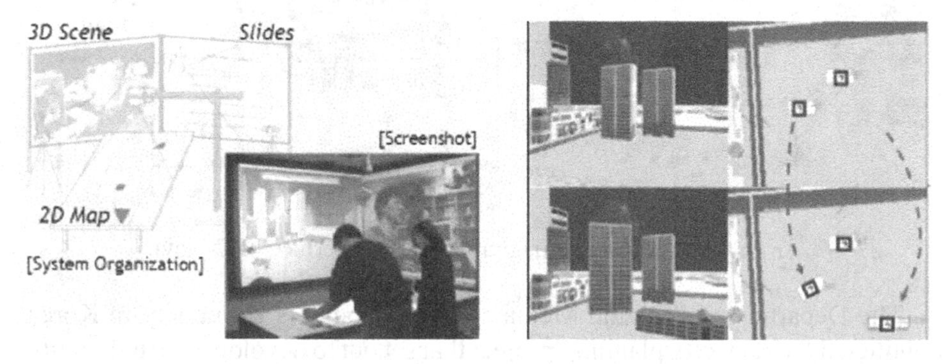

Figure 10. 2D maker-based tangible user interface of Virtual City Planning.

Figure 11. Scene displayed through an HMD (left) and, Interface of the hand-held computer (right). Computer Graphics and User Interfaces Lab, Columbia University, 1997).

Figure 12. 3D model mapped in a real world environment (left) and, Path-finding in the Desktop UI (right).

Bauhaus University also conducted a project called 'Laser Pointer Tracking in Projector-Augmented Architectural Environments,' which resulted in a new tracking system able to perform self-registration for interior environments. A tracking system detects a marker generated by a laser pointer using a camera that is able to move and rotate. The projector automatically analyses a shape and the reflexivity of objects and detects points generated by the laser pointer so that the projector automatically revises irregular surfaces and any spatial distortion. Through these methods, it is possible to test virtual objects in real environments, with the capacity for application to the design and investigation of buildings (Figure 13).

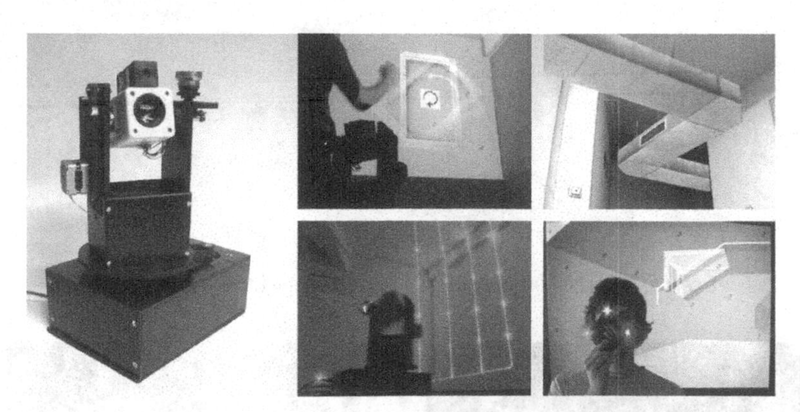

Figure 13. Laser pointer tracking system.

The HP New Media Lab in Portugal developed a system called MixDesign. In the latest version, users are able to develop a virtual model to match real objects so that architects can interact with real scale models using a tangible interface shaped like a paddle. The system cognises gestures with the paddle, which enables users to control some designs in augmented reality environments and to interact with virtual models in a real time manner (Figure 14).

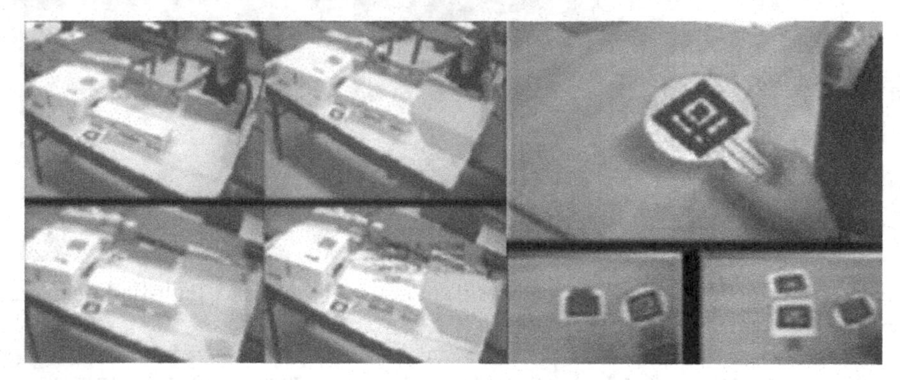

Figure 14. MixDesign on the table (right) and, Tangible interface (right).

Nancy Diniz and César Branco (from the Instituto Superior das Ciências do Trabalho e da Empresa in Lisbon) are developing a system called 'Touch me-Don't touch me!' that enables users to input virtual objects within augmented reality environments while web cameras detect motion and hand gestures utilising vision technology (Figure 15). Users wearing HMDs can interact with virtual object based movements that induce sounds and virtual effects (Figure 16).

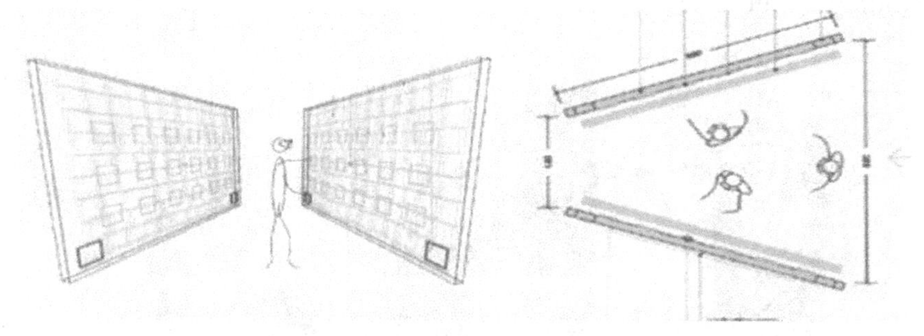

Figure 15. Interaction model of 'Touch me—Don't touch me!'

Figure 16. Virtual objects mapped in real worlds.

Architectural Anatomy is part of a project conducted by Columbia University (Feiner *et al.*, 1995) that explored the relationship between architecture and structural systems using augmented reality, virtual reality, and artificial intelligence. The building structure is displayed in a real building space based on augmented reality. This enables users wearing HMDs to see, not only detailed elements of the building like beams, columns, reinforcement et cetera, but also a commercialised analysis program of building structure (Figure 17).

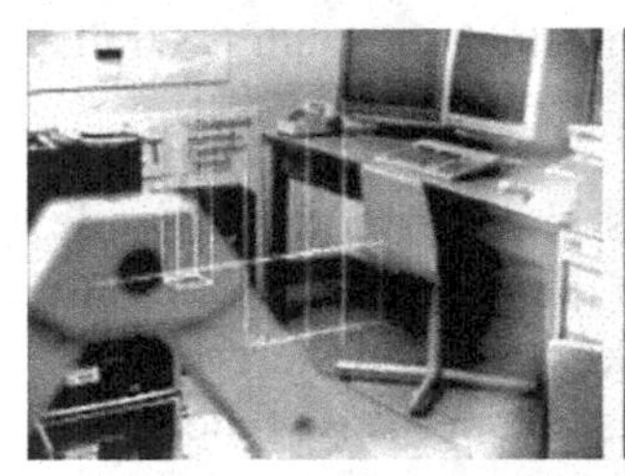
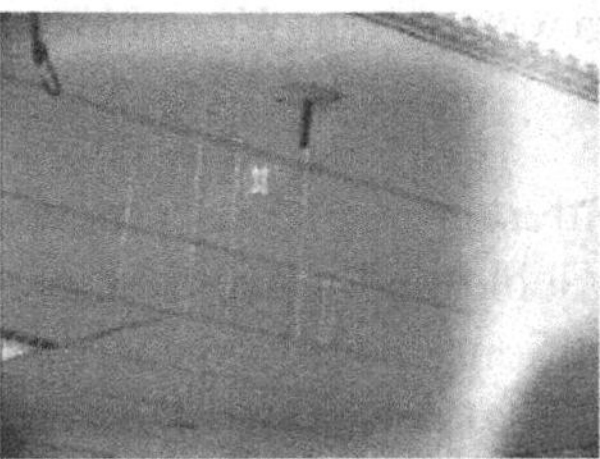

Figure 17. The Architectural Anatomy display: Computer Graphics & User Interfaces Lab, Colombia University, 1995 (Feiner *et al.*, 1995).

The Tinmith project team of the Wearable Computer Lab at the School of Computer and Information Science, University of South Australia developed the Tinmith-Metro AR system, which transforms real objects into a virtual 3D model using a set of instrumented pinch gloves as an interface. The system enables users with wearable computers to not only arrange some street objects like trees, tables, lighting et cetera, but also to edit and adjust them accurately. Furthermore, the system can make simple models more sophisticated by editing shapes and mapping textures. Thus, the system is effectively able to design landscape and give users the opportunity to gain prior experience before they physically construct landscapes and allows designers to suggest several designs for a specific potential building (Piekarski and Thomas, 2001) (Figure 18, 19).

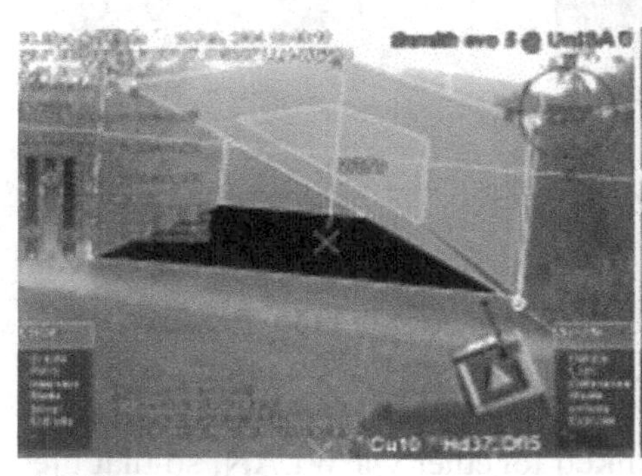

Figure 18. An example of object generation (Wearable Computer Lab, South Australia University, 2001).

Figure 19. Controlling 3D data with gloves (Wearable Computer Lab, South Australia University, 2001).

At the SIGGRAPH Conferences in Los Angeles in 1999, and in Rio de Janeiro in 2000, Steven Schkolne presented two new Surface Drawing Exhibitions. In these Exhibitions, people wearing head-tracked stereoscopic shutter glasses were able to begin creating shapes instantaneously. One basic method of developing surfaces utilised in that context was the automatic capture of a hand movement as a digital stroke. An instrumented glove detected each hand motion, which the computer recorded and displayed as a coherent stroke. The system can be used for the conceptual design of any 3D object, including buildings, characters, cars, clothes, furniture, and even roller coasters. The applications of surface drawing are currently being investigated in collaboration with an industrial design firm called Designworks/USA, who is working on the design of conceptual prototypes of products (Figure 20).

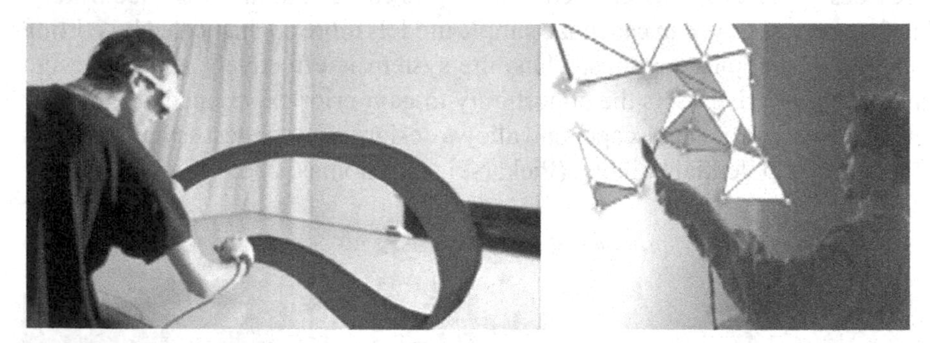

Figure 20. Surface drawing (left) and Crystal method (Schkolne, 2003) (right). Multi-res Modeling Group, Caltech, 2001.

The ARCHEOGUIDE, which stands for the Augmented Reality-based Cultural Heritage On-site Guide, produced the first mobile augmented reality guide for outdoor archaeological sites. ARCHEOGUIDE consists of a central server and a set of mobile devices all linked together via WLAN, so that the system can be employed at any archaeological site. Users who visit archaeological sites will be from a variety of educational backgrounds, age groups, and nationalities and have varying degrees of archaeological knowledge or computer skills. All of these users will be able to experience the system and its features through one of a range of different mobile devices available each of which caters to different preferences and styles of use (Figure 21).

Figure 21. ARCHEOGUIDE: Augmented reality-based cultural heritage on-site guide (Ioannidis, 2002).

An important aspect of ARCHEOGUIDE is the automatic selection of information adapted to the user profile. To achieve this, the server database objects are assigned to specific user profiles. Thus, once users enter their profile into the AR device, it automatically selects those items matching the user's profile and presents them in accordance with the user's position and orientation and the underlying rules and conditions associated with the data. This system gives individuals the freedom to interact in a user-friendly way that is simultaneously informative and pleasant (Figure 22).

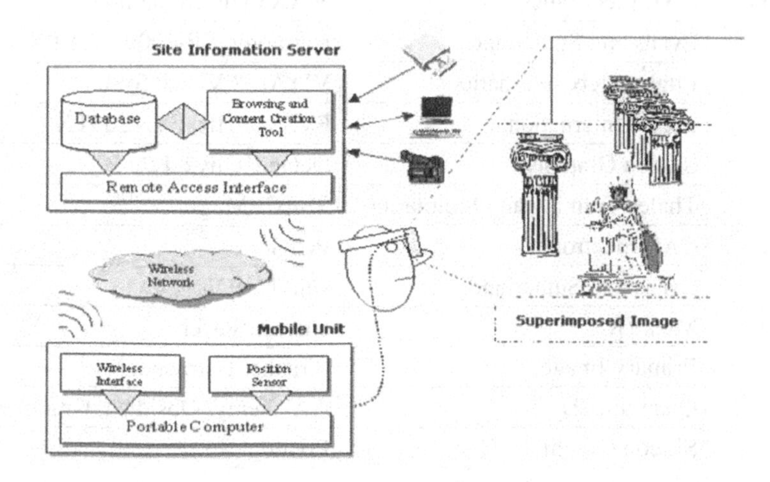

Figure 22. Schematic diagram of ARCHEOGUIDE (Ioannidis, 2002).

2.3. MARKET TRENDS FOR AUGMENTED REALITY

The solution market of augmented reality for supporting architectural design classifies it as a market primarily composed of image generators, visual simulation S/W and, tracking system H/W, S/W. An image generator is a computing system that combines information from multiple databases and generates a series of images based on graphics and rendering technology. In particular, image generators based on personal computing environments already feature on the market and this is gradually expected to expand into image generators based on mobile devices.

Visual simulation software includes many different types of software that apply to the simulation of the virtual environment, as well as to software that is used specifically for image generation. Visual simulation software is also often classified as object modelling/virtual prototyping software, as real-time 3D simulation software, and as geographic DB development software. Object modelling/virtual prototyping software is used for developing core components of visual simulation and, while real-time 3D simulation software enables the OS to simulate in a real-time manner; geographic DB development software is used for generating the background of virtual simulation.

2.3.1. The Market for Image Generators

As the PC-IG (Personal Computer-based Image Generator) market, which offers a low-end solution based on platforms such as Microsoft Windows NT and Linux, has been revitalise; companies related to PC-IG solutions have increased in number. This growth has resulted in the development of several

Table 1. Main vendors, products of IG.

	Vendor	Product
High-end IG	CAE Electronics	MAXVUE, Medallion
	Evans and Sutherland	Harmony, EP-1000CT, EPX
	FlightSafety International	VITAL 8 Visual System
	Frasca International	FVS 200HR, FVS 200TX
	Silicon Graphics	Octane, Onyx 4
	Thales Training and Simulation	SPACEMagic
PC-IG	CAE Electronics	PCIG
	Evans and Sutherland	simFUSION
	MetaVR	channelSurfer
	Primary Image	Piranha, Barracuda
	Quantum 3D	AAlchemy, Obsidian, Graphite
	Silicon Graphics	Prism

applications designed for military use, and both the military market and related general markets are expected to expand rapidly. Commercial solutions have been growing in the image generator (IG) market, which had previously been primarily occupied with military-specific solutions. According to Frost and Sullivan's presentation in 2001, commercial solutions occupied 36% of the IG market in 1997 and 42% of the IG market in 2000 (Lee, 2005).

2.3.2. The Market for Visual Simulation S/W

In previous years, most firms developing AR/MR systems imported the core technique from abroad then developed individual applications. Now, Korean domestic firms are largely developing new core techniques themselves in addition to expanding their applications. This increase in commercial solutions in the market for visual simulation may result in a market that is 74% occupied by commercial solutions by the end of 2007 (Lee, 2005). AR/MR technology has largely been developed for the military, although it has been used for commercial purposes as well, especially for the development of virtual museums, exhibitions, games, and other cultural and amusement projects.

2.3.3. The Market for Tracking and Matching H/W, S/W

The market for tracking and matching H/W, S/W is classified as marker based image matching and model-based image matching, according to the specific technology. There are several methods, like LEDs and flat markers, for performing marker-based image matching, as well as systems based on boundary-based matching and template matching, which relate to model-based image matching. Currently, the most prevalent method of tracking matches virtual objects by assigning specific points manually on a screen. Unfortunately, the accuracy of this method is limited when points are ambiguous or beyond the boundaries of the screen. Not surprisingly, research into hybrid tracking, which combines more than two methods of matching, in addition to supersonic waves and global positioning system (GPS) data is on the rise.

2.3.4. The Market for Mixed Reality Toolkits

Current popular products in the MR library and toolkit are shown in Table 2. In this table, the ARToolKit developed by HIT Lab has multi-platform support, consists of easy APIs, and has the widest international use (Lee, 2006).

Table 2. Toolkit, library of MR.

Areas	Products
Library	ARToolKit, MR-Platform, MxToolKit, ARTag, OSGART
Toolkit	AMIRE, APRIL, DART, ULTRA Authoring Tool, CMIL++

In December 2006, OSGART, updated from ARToolKit, which was presented and applied to a commercial product called MagicBook.

2.3.5. Results

As the market for AR solutions has become more active based on platforms such as AMIRE, APRIL, DWARF, Immersive Authoring, U Create, and CATOMI, the firms developing AR solutions have similarly increased in number. A drop in prices for image generators for visual simulation resulted in AR applications for architectural design becoming more diverse. Moreover, as realistic and productive visual simulation continues to develop, the market for architectural design systems based on AR is expected to grow rapidly. In the USA, the Boeing Corporation developed a wire assembling prototype system, while in Japan, in 1997, the Key Technology Research Center and Canon worked together to set up Mixed Reality Systems Laboratory Inc., a firm which develops 3D imaging and display equipment for MR.

As these examples show, technologically advanced countries like the United States and Japan are actively investing in augmented reality technology, which is a core technology for the new millennium. In addition, as a wider range of AR research becomes possible with the increasing shift of AR technology from indoor to outdoor applications, it is critical to develop practical solutions to allow this new and important technology to be used in the construction industry.

3. Behaviour Analysis Through an Analysis of the Life Cycle of the Construction Industry

3.1. DEFINITION AND LIMITS OF THE CONSTRUCTION INDUSTRY

The construction industry can be classified as both a construction business and as a construction service business. The designation of construction business includes those engaged in construction for environmental facilities; those providing services for industrial facilities (such as the construction and architecture of civil buildings, landscapes, bridges, harbours, roads, water supply and drainage sites, airports, urban areas, et cetera); plants and factories, facility management areas; and the construction and deconstruction of mechanical equipment. Construction service businesses are those that include surveying, design, supervision, business management, and maintenance management. According to this general definition and classification, it is possible to deduce in which areas AR technology can be integrated effectively.

3.2. ANALYSIS OF THE BUILDING LIFE-CYCLE: IN CONSTRUCTION

The construction industry classifies building life cycles into a series of operational steps including: the development of an operation plan, a project plan, site analysis, design (schematic design, design development, construction), occupancy, management, remodelling, and demolition.

During the operation plan, site analysis is conducted, taking into account the examination of economic efficiency and architectural building codes, before building type and range are estimated at the project plan stage. Next, site analysis takes place, concerning issues like sunlight, view, ventilation, transportation, et cetera, and a conceptual design is formalised. At the design stage, a detailed design is developed in addition to performance simulation. This requires trained workers and construction management (PERT-CPM) for the process to continue. After the occupancy stage is reached, the occupant and manager who support building maintenance play a main role. At the remodelling step, according to occupants' needs, it is possible to change the building type and improve building performance. At the end of the building life cycle, the building is demolished. The issues classified by each step are shown in Figure 23.

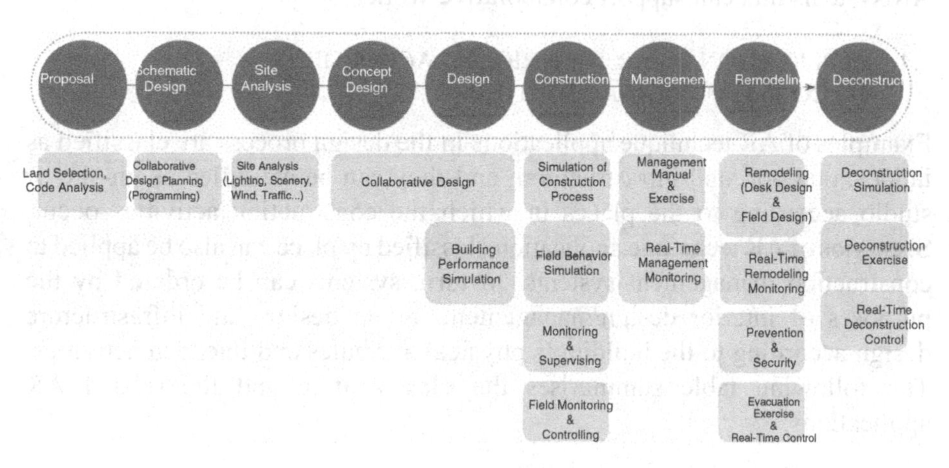

Figure 23. Architectural design process.

3.3. ISSUES ACCORDING TO PROCESS IN CONSTRUCTION INDUSTRY

3.3.1. One Core Domain of the Construction Industry

The construction industry consists of broad areas such as architecture and civil engineering involved with landscapes, bridges, harbours, roads, water supply and drainage facilities, and buildings such as airports, et cetera. However, these need to be classified more simply for the application of AR technology.

To address this problem, the first suggested classification is according to indoor and outdoor structures, and then to classify other items, for example, housing complexes, which are connected to civil structures and landscapes, roads, which are, in turn, associated with urban design and civil engineering, and, bridges, which link to industrial foundations.

3.3.2. Issues Classified by Industry Area/Main Participants

Participants concerned with construction are stakeholders (site owners, investors, and clients), architects, construction managers, construction engineers, users, and managers. According to a variety of domains, the main participants and their concrete activities can be classified by each step of the construction process. AR technology can be effectively applied at the stages of design, construction, and maintenance management. Real operators of these application domains are limited to contractors, managers, et cetera. Hence, a solution that is able to apply to the traditional design process and supports automation for the whole design process is necessary. Since applications of AR techniques to the design stage where decision-making relating to shape and surrounding surveys take place are still at a low level, it is necessary to develop AR systems that can support collaborative work.

3.4. APPLICATION OF AR TECHNIQUES ACCORDING TO CONSTRUCTION ACTIVITIES

Examples of AR technique applications in the design process are classified as indoor, outdoor, and infrastructural, and they can be classified as on-site or studio according to the places in which the construction activities occur. Scenarios of AR technique application classified by place can also be applied to construction management systems. In turn, systems can be ordered by the purposes of interior design/management, urban design, and infrastructure design according to the building's physical attributes and intended activities. The following table summarises the classification and the related AR applications.

4. Construction Management Systems Based on AR Techniques

4.1. REALISTIC AR TECHNIQUE BASED ON CONSTRUCTION INFORMATION

Based on the construction information model previously described, the 5 core techniques for realistic AR environments can be categorised as: a modelling technique that visualises physical forms, a tracking technique that visually integrates real environments with virtual objects, a rendering technique that

Table 3. AR application scenario.

Categories		Construction activities
Indoor	On-site	Application object: building indoor Application plan: design collaboration in real-time (material checking, decision-making), programming, remodelling
Indoor	Studio	Application object: building indoor/outdoor Application plan: conceptual design, remodelling, programming, simulation, building performance analysis
Outdoor	On-site	Application object: building outdoor Application plan: analysis of business efficiency based on GIS, negotiations, building scale estimation
Outdoor	Studio	Application object: building indoor/outdoor Application plan: site analysis, programming, review of design alternative, building shape analysis
Infra-structure	On-site	Application object: infrastructure (housing complex, road, bridge etc.) Application plan: site selection, decision-making, early construction plan, construction section plan, review of design alternatives
Infra-Structure	Studio	Application object: infrastructure (housing complex, road, bridge etc.) Application plan: site selection, decision-making, early construction plan, simulation, review of design alternatives

realistically generates geometry, and an authoring technique that generates a final output based on a database of construction information models. Moreover, more advanced realistic AR techniques are expected to be developed using sensors and motion tracking equipment. In order to apply this to a design support system based on construction information, the concept of a realistic AR technique is expressed in Figures 24 and 25.

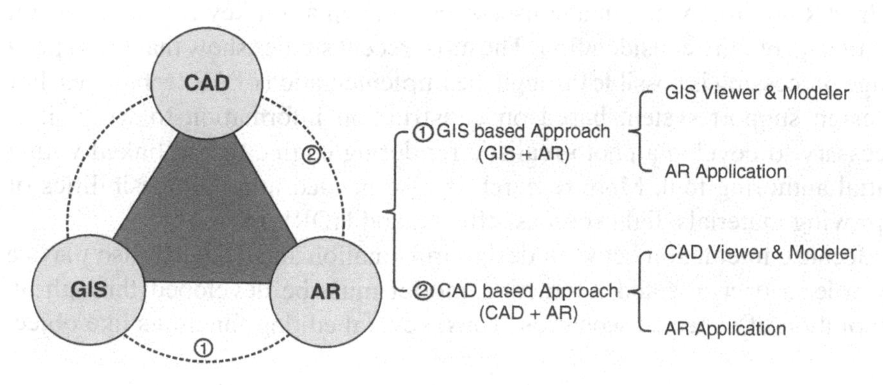

Figure 24. Concept of a construction system using AR.

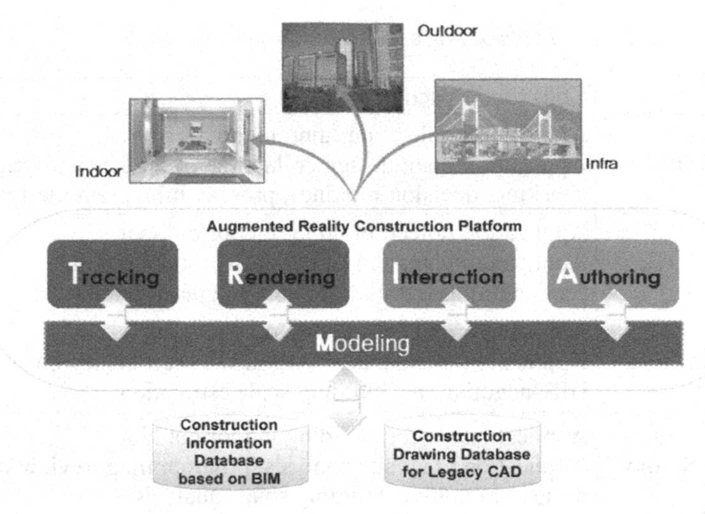

Figure 25. Concept of a construction system using AR.

The modelling technique will be actualised as an authoring tool for an object-oriented construction information model, which includes a management function for a spatial model using scripts and parameters that are able to generate object modelling – with a library created by the user. Through this system, the concept of 'Space Dynamics,' 'Programmable Space' is embodied and dynamic simulation is made possible according to colours and lighting.

In the past, tracking techniques manually detected specific points, which caused some problems such as inaccurate tracking, low speeds, and low efficiency. To address these problems, it is necessary to use a calibration method for mapping two worlds and to digitise and otherwise laborious and inaccurate method. To effect this, we first need to detect the edges and specific points of an object, so that it is then possible to calculate the correlation of detected points in 3D environments. Performing these tasks makes it possible to detect a coordinate system in the real world.

In AR environments, image user-experience plays a key role and several parameters require consideration. The most recent studies show that high speed image processing is possible through the implementation of AR techniques. For a design support system based on construction information to work, it is necessary to develop a photo-realistic rendering engine that is linked with a spatial authoring tool. More research is also needed into the possibilities of improving materials, light sources, effects, and HDRI images.

Because interaction between design information and designer also plays a key role, a user-oriented interaction model must be developed through an exploration of expected scenarios. Thus, several editing functions like object insertion, connection, rotation, movement, and scale adjustments like multi-modal interfaces are necessary so that users can control design in a real-time manner. Given the amount of research currently underway on interaction techniques that can detect users' gestures and movements without equipment, a highly advanced level of interaction can be expected.

An authoring tool based on the construction information model, which offers several methods of spatial authoring, could improve authoring techniques for AR. If a user is able to install an object in a design component database and edit a parameter of the object, then the work of design could become faster and more productive. Moreover, as users are able to upload and arrange multimedia in generated spaces, rich communication based on AR environments will become available.

4.2. AN IDEA OF ULTIMATE SYSTEM

By integrating modules of each technique, I have suggested some application systems that can be applied to interior design, urban design, and road and bridge design based on the development of a platform that will support it. Incidentally, it is also important to evaluate the efficiency of the design support system by conducting pilot projects (Figure 26).

For an interior design AR application system a designer would want to test building performance, analyse building shape, plan circulation, and analyse energy at the project plan stage, but also be able to work throughout the whole design process. In addition, the ideal system would support construction management and evacuation simulation as well.

For urban design, the AR application system would need to be able to conduct investigation and negotiation of business efficiency, and would estimate building scale, et cetera by linking with GIS. Particularly at the stage of schematic design, the system would have to act as a professional simulation tool to visually support design work. These functions would support visual expression for building façade design and landscape design so that designer's judgment could be examined. Additionally, a 5D simulation, which adds the concepts of schedules and security, could also be explored.

For infrastructure design, it is important to offer a high quality image to map onto the real world. By integrating GIS data, the system would need to enable the user not only to analyse construction section plans and to perform soil analysis, et cetera but would also offer additional services that support documentation of management manuals and location guides.

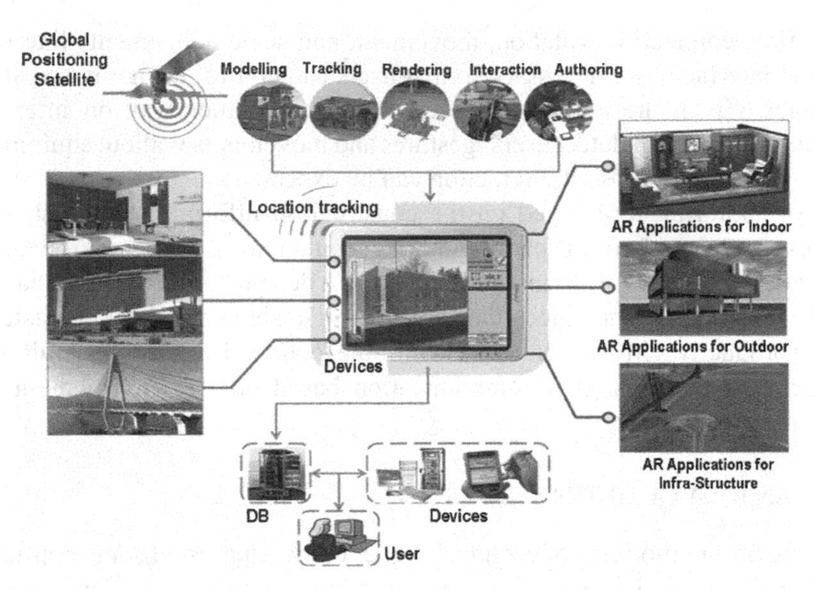

Figure 26. Design support system using realistic AR.

5. Conclusion and Discussion

A design support system using realistic AR techniques can be expected to cause an economic ripple and a shift of paradigm similar to that created by the appearance of the CAD/CAM technique. A variety of application services could become available through the combination of design automation systems and location-based mixed reality techniques, which should bring about a change in the IT industry directly, and should indirectly, have a significant impact on the traditional construction process. Accordingly, it would be desirable to offer AR application systems in broad domains, in order to support collaborative work in optimised environments.

EXPLORING PRESENCE AND PERFORMANCE IN MIXED REALITY-BASED DESIGN SPACE

XIANGYU WANG AND MI JEONG KIM
The University of Sydney, Australia

Abstract. This chapter presents an exploratory investigation on measuring the extent of presence in a MR-based design space through a comparative study using a tabletop system with two different types of displays: HMDs and 2D screens. This study explores the link between object presence and design performance in immersive MR-based design space while manipulating 3D blocks representing virtual furniture. A post self-reporting questionnaire was the main method of evaluation used. The results reveal that compared to the immersive HMD, the non-immersive 2D screen provided a more natural movement of objects and more realistic experience due to lower interface awareness, thereby improving design performance.

Keywords. Presence, Augmented Reality, Virtual Reality, Mixed Reality, Collaborative Design.

1. Introduction

In practice, designers render 3D images on 2D computer screens using CAD packages such as ArchiCAD and AutoCAD, which offer only limited interactions in design. There has recently been an increasing number of design applications that have adopted the concept of Mixed Reality (MR) to provide immersive or semi-immersive design spaces wherein real and virtual world objects are presented together on the one display (Daruwala, 2004; Dias *et al.*, 2002; Fjeld *et al.*, 1998; Kim, 2007; Lee *et al.*, 2003). The effectiveness of MR-based spaces has often been linked to the sense of presence reported by users of those spaces. Presence is defined as the subjective experience of being in one place, even when one is physically not there (Witmer and Singer, 1998). Current research on presence has focused primarily on the concept of personal presence, 'transportation' or 'being there' in the Virtual Worlds (VWs)

X. Wang and M.A. Schnabel (eds.), Mixed Reality in Architecture, Design and Construction, 75–90.

(Schubert *et al.*, 1999a; Schubert *et al.*, 2001; Witmer and Singer, 1998), in which navigation and knowledge acquisition have been regarded as important tasks performed within virtual environments (VEs) (Satalich, 1995; Witmer *et al.*, 1996). Navigation is an inherently cognitive process and an understanding of the knowledge it requires can inform the study of navigation (Nash *et al.*, 2000). The most important task supported in MR-based design spaces may very well be 3D configurations to create or manipulate design objects, in which object presence, the subjective experience that a particular object exists in a user's environment, (Stevens and Jerrams-Smith, 2001) would be more critical for effective design performance.

This chapter presents an exploratory study on presence in an MR-based design space comprising a tabletop system with 3D blocks in ARToolKit. The MR tabletop system with 3D blocks can provide intuitive interactions with virtual designs and immersive 3D visualisation. Many researchers have assumed the benefits of immersive visualisation for task performance (Arns *et al.*, 1999; Gruchalla, 2004). As Schuemie *et al.* (2001) stated, presence can be effective in itself for certain applications such as games and movies. Whether presence embodied by immersive visualisation can improve design performance is unknown, since there is no significant evidence that supports it. This study explores presence and task performance in immersive and non-immersive MR-based design space for a 3D modelling task. A post-session questionnaire was used for evaluative purposes, as subjective rating scale questionnaires have been widely accepted as the main measurement for studying presence (Lessiter *et al.*, 2000; Lombard *et al.*, 2000; Schuemie *et al.*, 2001; Slater and Usoh, 1993).

2. Research on Presence in Immersive Spaces

Research on presence in VEs is closely associated with psychological research, as the experience of presence is chiefly a mental state (Schubert *et al.*, 1999a). Most research on presence has attempted to understand the nature of presence; the factors that contribute to it, and the manner in which it is possible to measure it. Our study also starts with a synthesis of the literature defining presence.

2.1. THE CONCEPT OF PRESENCE

In order to understand the concept of presence, the concept of immersion should be first understood, as the two concepts are strongly related. This chapter adopts the work of Slater and Wilbur (1997), who define immersion as an objective description of display technology, such as stimuli from reality, a range of sensory modalities, a field of view and a display resolution. That is, 'immersion' represents what the system delivers to users, and thus can be objectively assessed, whereas presence is a psychological reaction to such immersion, and thus a subjective measure is mainly used to evaluate it (Nash *et al.*, 2000). Lombard and Ditton (1997) characterised the concept of presence using six features; social richness, realism, transportation, immersion, a social actor within a medium, and a medium as a social actor. Realism and transportation are the most popular features of the concept of presence proposed by researchers. 'Realism' represents the extent to which VW appears to be realistic and 'transportation' represents peoples' perception of being present in the VW. Schloerb (1995) proposed that objective presence and the likelihood of successfully completing a task, are more important criteria for a VE, compared to subjective presence – perceiving oneself to be physically present in the VE. Heeter (1992) distinguished three different types of presence; personal presence, social presence and environmental presence. 'Personal presence' refers to an ability to inhabit a VE remote from the real world (Slater and Usoh, 1993), giving the perception of the self being at a certain location. Many VEs provide avatars to create a strong sense of personal presence (Heeter, 1992) since the distinction between self and non-self occurs at the boundary of the body (Loomis, 1992). 'Social presence' is a sense of interacting with others, which can provide evidence for a user's existence, thus increasing presence (Heeter, 1992) and influencing a user's attitude (Heeter, 1992; Sheridan, 1992). In a similar manner to social presence, environmental presence refers to the reaction of the environment to the user, which can provide evidence of a user's existence (Heeter, 1992). The user's ability to modify the environment is an important factor in achieving environmental presence (Sheridan, 1992).

We consider object presence as one of the most important types of presence in the MR-based design space, since design performance is mainly carried out using 3D objects. Object presence is the ability to interact with an object, which can be enhanced either by isolating the object from reality or by providing a display that is capable of accommodating external events (Stevens *et al.*, 2002). That is, the object and distracting event can exist in the same reality due to the seamless boundary between a VE and the real world. However, in terms of object presence, the ability to become 'involved' may not be easily destroyed by external events since a user usually depends on their past experience of how to interact with the display (Stevens *et al.*, 2002). Object presence is often thought to be linked to scene depth for close object viewing (Rokita, 1996), and would be supported by a high resolution display, a wide field of view and the addition of audio and haptic information (Sheridan, 1992).

Schuemie *et al.* (2001) described several theories on the nature of presence in terms of underlying factors; presence as non-mediation, exclusive presence, presence by involvement, ecological view, social/cultural view, estimation theory and embodied presence. Our study adopts the embodied cognition framework for the analysis of presence. Embodied cognition emphasises the formative role the environment plays in the development of cognitive processes when a tightly coupled system emerges from real-time interactions between organisms and their environment. One necessary condition for cognition is embodiment, which is understood as the unique way an organism's sensorimotor capacities enable it to successfully interact with its environment (Cowart, 2006). In the same context, Schubert *et al.* (1999a) proposed embodied presence; presence which emerges from interactions with an environment as the possible bodily actions in the VW. That is, VEs are mentally represented as meshed patterns of actions and that presence is experienced when these actions include the perceived possibility of navigating the body or manipulating objects in the VE. They argued that presence should involve two components: the suppression of stimuli from the real environment and the mental construction of a space out of the VE in which the body can be moved. In other words, conflicting projectable features from the real world must be suppressed for presence to emerge and the VE must be perceived of in terms of embodied action.

2.2. MEASURING PRESENCE USING QUESTIONNAIRES

Given that presence is a mental state, the most commonly used measurement method for presence is the questionnaire that directly asks about participants' sense of presence through post-test ratings (Nash *et al.*, 2000; Schuemie *et al.*, 2001). A number of subjective questionnaires have been developed, in which researchers determined several major components of presence based on certain theories (Lessiter *et al.*, 2000; Lombard *et al.*, 2000; Schuemie *et al.*, 2001; Slater and Usoh, 1993). Through analysing the questionnaires, researchers have attempted to find several factors on presence using statistical techniques such as factor analysis (Kim and Biocca, 1997; Schubert *et al.*, 1999b; Witmer and Singer, 1998). The results of the questionnaires often provide feedback that elucidates the understanding of presence, thus leading to the refinement of the presence theories on which they are based (Schuemie *et al.*, 2001). The following three questionnaires have received much attention and have been utilised by many researchers; Slater and colleagues' questionnaires (Usoh *et al.*, 2000); the Presence Questionnaire (PQ) and the Immersive Tendencies Questionnaire (ITQ) (Witmer and Singer, 1998); and Igroup Presence Questionnaire (IPQ) (Schubert *et al.*, 1999b). Many researchers have developed their own customised questionnaires.

Slater and colleagues' questionnaires (Slater, 1999; Usoh *et al.*, 2000) are based on three presence indicators: the subjective sense of 'being there' depicted by the virtual display; the extent to which the VE becomes more 'real' than everyday reality; and the sense of having visited a 'place' rather than having seen images. Witmer and Singer (1998) identified involvement and immersion as conditions for presence and determined several factors that influence a sense of presence: Control factors – the amount of control the user had on events in the VE, Sensory factors – the quality, number and consistency of displays, Distraction factors – the degree of distraction by objects and events in the real world, and Realism factors – the degree of realism of the portrayed VE. Their Presence Questionnaire (PQ) was designed based on the above factors, which attempt to measure the causes of presence, as evaluated by the user. In addition, they developed an Immersive Tendency Questionnaire (ITQ) to measure differences in the capability or tendency of individuals to experience presence (Witmer and Singer, 1998). The ITQ has been shown to be statistically reliable and valid and to correlate with the PQ score for a number of studies. Schubert *et al.* (1999b) developed their Igroup Presence Questionnaire (IPQ) based on the construction of a spatial-functional mental model of VE. For the question items, they combined previously published questionnaires (Slater *et al.*, 1995; Usoh *et al.*, 2000; Witmer and Singer, 1998) with a questionnaire from earlier research and some newly developed questions. From a factor analysis, three presence factors and five immersion factors were extracted. The presence factors were spatial presence (SP), involvement (INV), and realness (REAL). The first two factors, spatial presence (SP) and involvement (INV) support the distinction between a spatial-constructive and the attention component, which was derived from the embodied presence model (Schuemie *et al.*, 2001). The immersion factors were Quality of Immersion (QI), Drama (DRAMA), Interface Awareness (IA), Exploration of VE (EXPL), and Predictability (PRED).

3. Presence Questionnaire in an MR-Based Design Space

A questionnaire was developed for measuring presence in the MR-based design space. The questionnaire comprises four categories: personal and object presence, involvement, interface and realness (Schubert *et al.*, 1999b). The 'interface' category representing the technical side of the VE was added for assessing the properties of the interaction between user and VE. This study is especially concerned with the affect of a variety of interfaces on presence. We do not include social presence or co-presence items in this study because the focus is on individual/solitary design work. Relevant questionnaire items were constructed based on theory and previously published questionnaires by other authors.

In terms of spatial presence, the focus is on object presence, ability to perceive locations of other objects, rather than on personal presence and the ability to perceive one's location. Thus, the questions for object presence were composed on the following items: the sense of an object being present or moving in the Augmented Reality (AR) environment; the sense of the object's tangibility; and the sense of ease with which the object can be examined or manipulated. All other questions included in this category are in Table 1.

Table 1. Questions in the 'presence' category.

Please rate your experience for each question on scale of 1–5 where 1 = none, 2 = poor/mild, 3 = moderate, 4 = good and 5 = excellent:	
1.	How **strong** was your sense of **being present** in the AR environment?
2.	How **strong** was your sense of **objects being present** in the AR environment?
3.	How **strong** was your sense of **objects moving** in the AR environment?
4.	How **strong** did it seem **as if you could reach out and touch the objects** in the environment?
Please rate your experience for each question on scale of 1–5 where 1 = not at all, 2 = a little, 3 = moderate, 4 = much and 5 = very much:	
5.	How **well** could you actively **examine** objects **via tangible markers**?
6.	How **well** could you actively **examine** objects **using vision**?
7.	How **well** could you **examine** objects **from changing viewpoints** on your own?
8.	How **well** could you **move or manipulate** objects in the environment?
9.	How **closely** could you **examine** objects?

The second category, involvement (see Table 2), is concerned with the allocation of attentional resources. Witmer and Singer (1998) defined 'involvement' as a psychological state experienced as a consequence of focusing one's attention on a coherent set of stimuli or related activities and events. Witmer and Singer posited that a person who concentrated on a VE would become more involved and as a result, would experience a higher sense of presence (Schuemie *et al.*, 2001). Several items that relate to the suppression and forgetting of the real environment are also included in this category since a high sense of presence in a VE requires a simultaneously low level of presence in the real world (Slater *et al.*, 1994).

Table 2. Questions in the 'involvement' category.

Please rate your experience for each question on scale of 1–5 where 1 = none, 2 = poor/mild, 3 = moderate, 4 = good and 5 = excellent:
10. How **strong** was your sense of **being involved** in the visual environment?
11. How **strong** was your sense of **being involved in the experimental task**?
12. How **strong** was your sense **of events occurring in the real world** around you while involved in the environment?
13. How **strong** was your sense of wearing **the HMD** or looking at **2D screen**?
14. How **strong** did you feel **comfortable** inside the environment?
15. How **strong** did you feel **entertainment** in the environment?
16. How **quickly** did you **adjust to** the AR environment experience?
17. How **proficient in interacting with** the AR environment did you feel?
18. To what extent did you feel **confused or disoriented** at the beginning of breaks or at the end of the experimental session?

The 'interface' category includes interface awareness; items that distract from the VE experience and the sensory quality of the environment, describing the richness and consistency of the multimodal presentation (see Table 3 for questions).

Table 3. Questions in the 'interface' category.

Please rate your experience for each question on scale of 1–5 where 1 = not at all, 2 = a little, 3 = moderate, 4 = much and 5 = very much:
19. To what extent did **the control devices distract** you from performing assigned tasks?
20. To what extent did **the visual display (HMD or Screen) distract** you from performing assigned tasks?
21. How **much effort** did your spend when wearing **the HMD** or looking at **2D screen**?
22. To what extent did **the lag** or **delay** between your actions and the response **distract** you from performing assigned tasks?
23. How **responsive** was **the AR environment** to actions that you performed?
24. To what extent did you find **easy to manipulate the virtual display**?
25. To what extent did you find **easy to look around the AR environment**?

The category 'realness' is a subjective way of measuring the perceived reality of the VE. Although it is not part of the actual experience of presence (Schubert *et al.*, 1999a), realistic reactions and emotions are one of the most important consequences of presence as shown in research on the relationship between presence and fear of heights (North *et al.*, 1998; Regenbrecht *et al.*, 1998). Table 4 listed all the questions for this category.

Table 4. Questions in the 'realness' category.

Please rate your experience for each question on scale of 1–5 where 1 = not at all, 2 = a little, 3 = moderate, 4 = much and 5 = very much:
26. To what extent did **the environment** seem **realistic** to you?
27. To what extent did **your movements** in the AR environment seem **natural** to you?
28. To what extent did **the depth and volume of the object** seem **geometrically correct,** that is, the right size and distance in relation to yourself and other objects?
29. To what extent did **the mechanisms** which controlled your movements in the environment **seem natural** to you?
30. To what extent did **your experience** of the environment seem **consistent with** your real world experience?
31. To what extent did **the environment's reactions** to your action seem **realistic**?
32. To what extent did you **anticipate** what would happen next in response to the actions that you performed?

The questionnaires also included an open question to elicit additional comments and several questions related to the participants' demographic details and background. The demographic information included gender, age, degree major, year of degree completion, and the number of years of playing Virtual Reality computer games.

4. Exploratory Study

A series of studies involving students as participants was conducted to analyse presence in an MR-based design space. Two different display conditions were devised, 2D screen and Head Mounted Display (HMD), in order to see if the immersive HMD influenced a user's subjective sense of presence, thereby affecting task performance in designing, compared to the non-immersive 2D screen. For a between-subject design, five participants were assigned to each condition, where participants had to solve a design problem. The reason that the between-subject design was chosen, was to avoid learning effects from repeated participation in a within-subject design. For example, prior experience in rating stimuli can affect the subsequent rating of presence (Freeman *et al.*, 1999) or the order in which the VEs are presented can influence the relative presence ratings of the VEs (Welch *et al.*, 1996).

4.1. STUDY SET-UPS: 2D SCREEN VS. HMD

The two study set-ups, using 2D screens and HMDs, were constructed with the same tabletop system with 3D blocks as devised by Kim (2007). The tabletop system includes a horizontal table and a vertical screen as shown in Figure 1a. 3D blocks representing furniture were made of square pieces of plywood, each with its own tracking markers made in ARToolKit (Figure 1b, c). ARToolKit is a free AR software using methods from computer vision, so that a web camera captures the patterns, and outputs the corresponding digital images on a vertical LCD display in real time. When users manipulate multiple blocks, each block allows direct control of virtual objects and provides tactile feedback. For the convenience of recognising a furniture model, its corresponding digital model image was shown on the side of the 3D block.

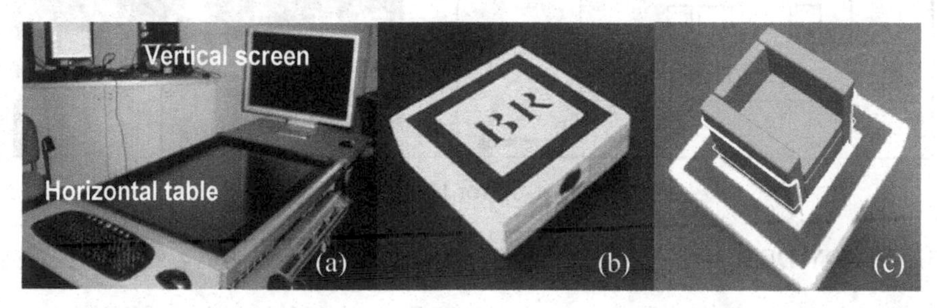

Figure 1. Tabletop system (a), 3D block (b), and digital model (c).

4.2. 2D SCREENS SESSION

Figure 2 shows the equipment set-up for the 2D screen session. This includes the tabletop system comprising a horizontal table, a vertical LCD screen, 3D blocks and a web camera. A 2D studio plan for the design task was placed on the horizontal table, on which some of the 3D blocks were initially located. The other 3D blocks, that be rearranged to any location, were assembled towards the

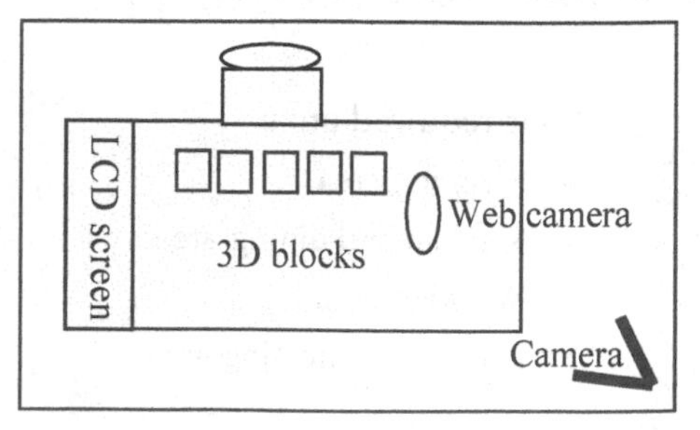

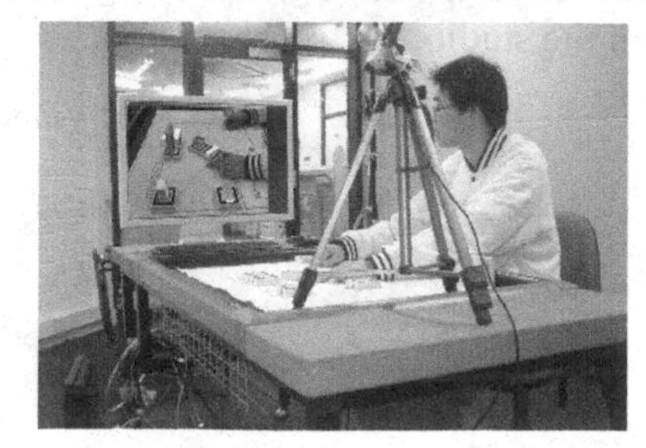

Figure 2. Set-up for 2D screen session.

front of the table. The web camera was set to the opposite side of the screen at a suitable height and angle to detect all the tracking markers of the 3D blocks. The screen is capable of up to 1600 × 1280 stereo resolution; however, due to HMD hardware constraints, the resolution used for this study was limited to 800 × 600.

4.2.1. HMD Session

Figure 3 shows the set-up for the HMD session. The overall set-up was similar to that of the 2D screen session. Instead of the vertical screen, a Head Mounted Display, hi-Res800^{TM}$_{PC}$Headset was used to project 3D digital images at a resolution of 800 × 600 with ±26.5 degree diagonal of field of view. Thus, virtual images were overlaid directly on a view of the real world using a see-through HMD.

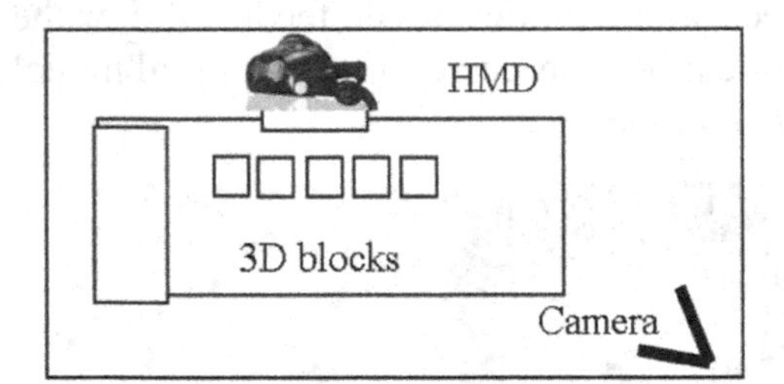

Figure 3. Set-up for HMD session.

4.3. STUDY DESIGN: PARTICIPANTS AND TASKS

Ten undergraduate students were recruited from the University of Sydney for the studies. They were of a similar age range (21–27 years) and educational background (design computing or digital media). Not all of them had played Virtual Reality (VR) computer games, but all were computer literate. We adopted the same space-planning task used by Kim (2007) since the framework seemed to be the most relevant for our study settings. The design task asked the subjects to re-design a residential studio into a home office as shown in Figure 4. All the subjects had to define four required areas: sleeping; kitchen and dining, working, and living and meeting, by arranging 3D furniture objects in the studio.

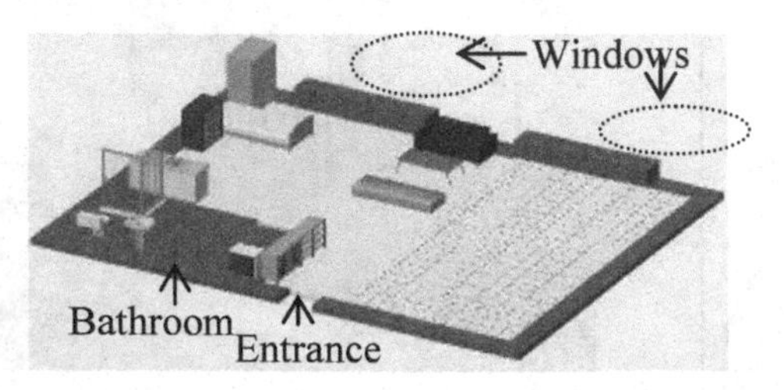

Four required areas:

- Sleeping area
- Kitchen and dining area
- Working area
- Living and meeting area

Figure 4. Home office task.

4.4. PROCEDURE

Although the manipulation of the 3D blocks is intuitive and natural, participants were engaged in training sessions prior to the study because 3D blocks and HMDs are unfamiliar input and output devices. They were instructed not to entirely block the markers on the 3D blocks from the web camera otherwise; the virtual model would either not display at all or flicker in and out. They participated in the studies for about 10–15 min after reading through the design brief. There was no time limitation and no special instructions regarding their verbalisation given to them. All design sessions were recorded on the video camera. After the session, they were asked to complete a post-test questionnaire.

5. Results

This section presents the preliminary results of this study, which includes statistical performance measurement and analysis of the questionnaires. We performed a Mann-Whitney U test on each item and category of questions, to examine significant differences between the 2D screen and HMD conditions. In addition, correlations were investigated among design performances, the total score of presence and question items in order to identify several factors influencing presence in the non-immersive and immersive environments.

5.1. RELIABILITY OF QUESTIONNAIRE' CATEGORIES AND ITEMS

First, assessing the post-test questionnaire's scale for reliability, a Cronbach's alpha test for internal consistency was conducted. The generated score was 0.88 ($N = 10$), which indicates the reliability of the questionnaire. Prior to the investigation of the differences between the 2D screen and the HMD conditions, we conducted a Pearson correlation test between the total score for the questionnaire and the score for each question category and item. All four categories correlated with the total score: 'object presence' ($r = 0.881$, $p < 0.05$), 'involvement' ($r = 0.711$, $p < 0.05$), 'interface' ($r = 0.754$, $p < 0.05$) and 'realness' ($r = 0.829$, $p < 0.05$). Specifically, the item in the 'object presence' category: 'How strong was your sense of objects moving in the AR environment?,' has a significant correlation with the total score of 'presence' ($r = 0.838$, $p < 0.05$). In the 'involvement' category, the items: 'How strong was your sense of being involved in the visual environment?' ($r = 0.645$, $p < 0.05$) and 'How strongly did you feel entertainment in the environment? ($r = 0.742$, $p < 0.05$)' showed significant correlations with the total score of 'presence.' Significant correlations were also found for the two items of the 'interface' category: 'How responsive was the AR environment to actions that you performed?' ($r = 0.733$, $p < 0.05$),'and 'To what extent did you find it easy to manipulate the virtual display?' ($r = 0.758$, $p < 0.05$). The item in the 'realness' category: 'To what extent did your experience of the environment seem consistent with your real world experience?' ($r = 0.633$, $p < 0.05$), also revealed a significant correlation with the total score of 'presence' questionnaire. These results indicate that the four categories may reflect the embodied presence as effectively as intended, in which the six items: 'object movement,' 'involvement in VE,' 'being entertained,' 'responsiveness,' 'ease manipulation,' and 'realistic experience,' are more closely associated with the embodied presence.

5.2. PERFORMANCE MEASURES

Figure 5 shows examples of the two conditions. The right one defined the four required areas including well-designed circulation and privacy in the studio whereas the left layout is missing the dining area. The increased time spent in the 2D screen session indicates that participants were more involved in designing than for the HMD condition. This improvement could be related to participants' object presence in the environments, because the screen had a much higher resolution than did the HMD.

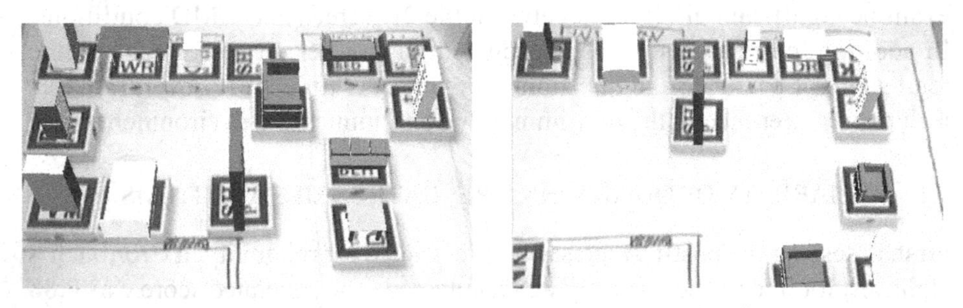

Figure 5. 2D screen session (a) and HMD session (b).

A Pearson correlation was calculated between performance and each category of the questionnaire in order to investigate factors contributing to design performance. Significant correlations were found in the two question items in the 'realness' category: 'To what extent did your movements in the AR environment seem natural to you?' ($r = 0.888$, $p < 0.05$), and 'To what extent did your experience of the environment seem consistent with your real world experience?' ($r = 0.647$, $p < 0.05$). These results indicate that design performance is more closely related to natural object movements and realistic experience in the design space. As expected, the scores on these two items turned out to be higher for the 2D screen session in the analysis of the questionnaire.

5.3. ANALYSIS OF THE QUESTIONNAIRE

The following figures show mean scores of all items in the four categories between the 2D screen and HMD sessions. For consistency of analysis, the scores of three items in the 'involvement' category and four items in the 'interface' category were recalculated reversely and indicated with ® because they were assumed to have negative relations with the 'total presence score.' Accordingly, the greater the score of the item, the more it is positively associated with the 'total presence score.'

Firstly, we investigated which items produced different scores between two design sessions. As shown in Figure 6, the items of the 'object presence' category have a similar distribution of scores for the two sessions. Thus, there was no significant difference among the items. Conversely, we found a significant statistical difference in an item in the 'involvement' category: 'How strong was your sense of events occurring in the real world around you while involved in the environment?' ('involvement' category, Z = -147, N = 10, p < 0.05). As explained above, this 'awareness of event' item indicated by ® should be interpreted inversely because it was assumed to be negatively related to the 'total presence score.' This finding indicates that participants in the 2D screen session tended to sense more events that distracted from outside whereas the immersive aspect of the HMD may protect participants from external disturbances. Although there were no other significant statistical differences, scores of the items in the 2D screen session, except one item 'awareness of display,' reflect designers' greater involvement in the design environment, as shown in Figure 7.

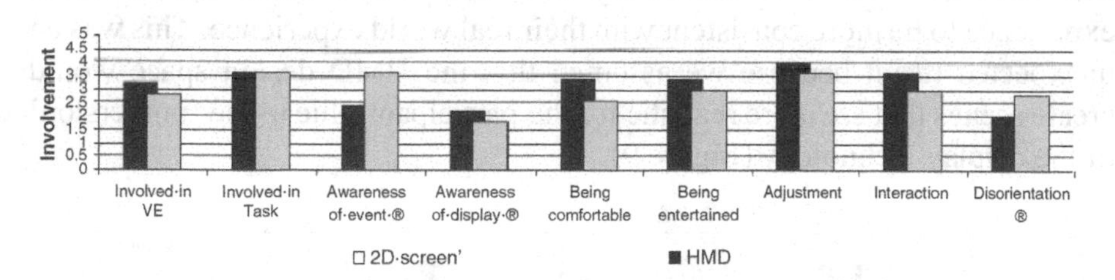

Figure 6. Mean scores of the items in the 'object presence' category.

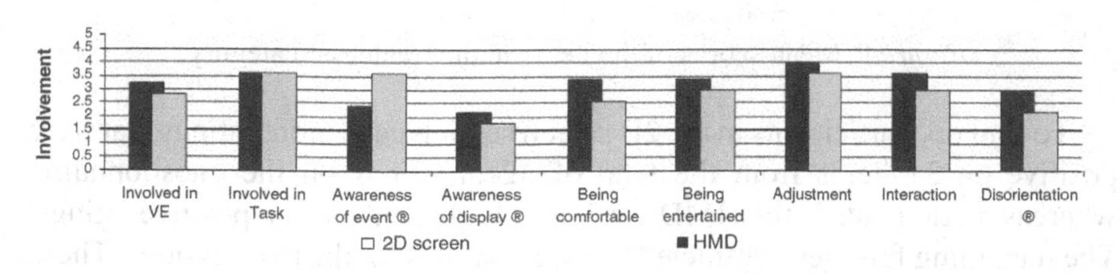

Figure 7. Mean scores of the items in the 'Involvement' category.

In terms of the 'interface' category, we found significant differences in two items: 'To what extent did the HMD or Screen distract you from performing assigned tasks?' ('interface' category, $Z = 1.735, N = 10, p = 0.083$), and 'How much effort did your spend when wearing the HMD or looking at 2D screen?' ('interface' category, $Z = -2.081, N = 10, p < 0.05$). These results suggest that distractions, and effort expended in relation to the HMD are greater than those caused by the 2D screen. As shown in Figure 8, all items of the category 'interface' showed that designers were more aware of interfaces in the HMD session, being distracted from the MR-based design experience.

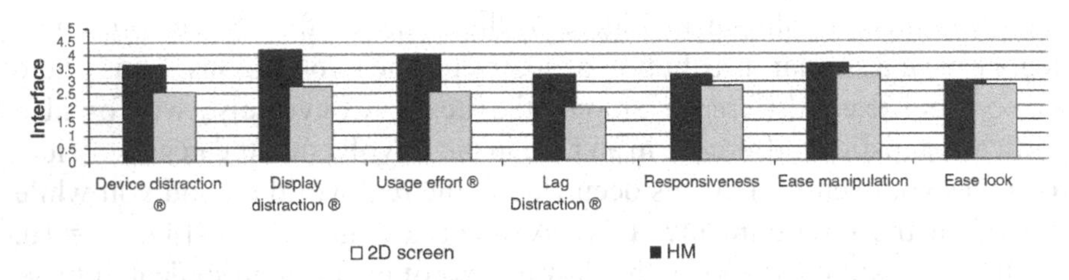

Figure 8. Mean scores of the items in the 'Interface' category.

The last two items that showed significant differences between the two sessions belong to the 'realness' category: 'To what extent did your movements in the AR environment seem natural to you?' ('realness' category, $Z = -1.928, N = 10, p = 0.054$), and 'To what extent did your experience of the environment seem consistent with your real world experience?' ('realness' category, $Z = -2.739, N = 10, p = 0.06$). Surprisingly, participants in the 2D screen session seemed to sense the movement of objects as being more natural and found the experience to be more consistent with their real world experience. This was an unexpected result because we assumed that the HMD design space would create scenes that are more realistic for the participants due to the 'immersion' of the display technology (Figure 9).

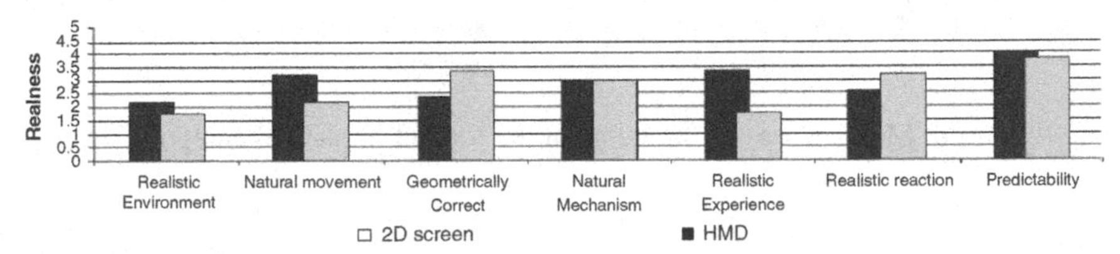

Figure 9. Mean scores of the items in the 'realness' category.

To sum up, participants in the 2D screen session gave more ratings that were positive on 21 items from the total of 32 questions on the questionnaire, whereas participants in the HMD session provided only seven positive ratings. The remaining four items attracted the same ratings for the two sessions. These results suggest that the HMD did not create the immersive space as intended, thereby causing participants to experience unnatural MR as compared to the non-immersive 2D screen. In the final analysis of the questionnaire, we used statistics to compare the means of each category and the total presence score in order to investigate the difference in the category level rather than in the item level. There was a significant difference in the 'interface' category ($Z = -2.147$, $N = 10$, $p < 0.05$) between the two sessions, in which the lower the score, the more the participants were aware of the interface. This result indicates that compared to using a 2D screen, participants wearing HMDs were more aware of interfaces distracting from their sense of presence and experience in the space.

5.4. PARTICIPANT COMMENTS

Participants' comments on the 2D screen session reflect two aspects of problems relating to 3D blocks in ARToolKit, rather than aspects of the 2D screen itself. They complained about the unstable visualisation, such as flickering of pieces and lag time of the webcam. In addition, in some cases where the objects were smaller than the blocks, it was hard to fit furniture exactly, thus making it difficult to design the spatial layout in a realistic manner. Similarly, participants in the HMD session also made comments on equipment limitations. These participants complained about the distractions caused by incorrectly rendered images, such as blurry and low-resolution images in the HMD. Some participants said that the lag-time between moving objects and their hand caused confusion, so that they had to rely on their sense of touch rather than vision in order to navigate accurately through the VW. Further, although easier to manipulate, the movable object got in the way of placing and orienting pieces to fit next to other pieces since it was difficult to avoid 'covering' up the symbol when picking up the markers. The largest distraction associated with 'being immersed' in the environment was a feeling of discomfort in the HMD. Participants reported that it felt as if the HMD was going to fall off, so they were distracted by the need to ensure that it stayed on. In addition, the limited field of vision which the HMD imposed made them feel quite confined, and hindered their ability to determine spatial dimensions and locations.

6. Discussion and Conclusion

The results of this study reveal that the non-immersive 2D screen condition provided object movements that are more natural and a more realistic experience, thereby improving design performance, when compared to the immersive HMD condition. From the total of 32 question items, participants in the 2D session responded more positively on 21 items. The unstable visualisation of the objects seemed to be more severe in the HMD session, which might increase the disturbance of participants' involvement in the immersive design space. Primarily, the uncomfortable feeling in the HMD destroyed much of participants' sense of presence. That is, the immersive design space created by the HMD did not support participants' design performance effectively, thereby raising questions about its effectiveness as a design tool. Accordingly, the assumption that the added value of immersion would greatly benefit design performance is uncertain based on these results.

The goal of this study was to identify the performance and sense of 'object presence' in the HMD condition when compared to the 2D screen condition, especially for the activity of designing in MR-based space. In general, this study is insufficient to determine which factors contribute to design performance, since immersion alone is not a sufficient construct. However, this empirical study has provided more insight into which aspects of the MR systems should be considered to develop MR design systems. Contrary to our expectation, the results show that non-immersive displays produced better design outputs by offering natural movements and realistic stimuli to the participants compared to the immersive display. We used only a subjective self-reporting questionnaire for measuring object presence, involvement, interface awareness and realness, which might rely on a personal interpretation of the question. If participants fail to understand the relationship between a question and their experience, they could generate an incorrect score (Freeman *et al.*, 1999). Thus, we will consider objective measurement methods for the next stage of this study. Further study regarding the development of the questionnaire measuring object presence will be carried out to investigate contributing factors to the design performance. In this study we focused on the visual aspects of the MR-based design space, however, given that the participants relied on their sense of touch rather than vision, the tangible aspects of interaction will be explored to further our understanding of design performance in MR-based design space.

3 MIXED REALITY IN ARCHITECTURE

MOBILE ARCHITECTURAL AUGMENTED REALITY

MARK BILLINGHURST AND ANDERS HENRYSSON
University of Canterbury, New Zealand

Abstract. Mobile devices provide a new platform for experiencing Augmented Reality (AR) architectural applications, and enable novel types of applications to be developed. In this chapter we review previous work in the area, suggest how it could be applied in an architectural setting, and describe promising future research directions. As AR technology is migrating to mobile phones exploring how these devices can be used to support architectural applications is increasingly important.

Keywords. Architecture, Augmented Reality, Mobile Phones.

1. Introduction

The traditional view of Augmented Reality (AR) is that it involves the seamless overlay of digital imagery onto the real world (Azuma, 1997). Over the past four decades AR research has focused on developing the basic enabling technologies (tracking, display, input devices) and exploring how these technologies can be used to enhance an individual's experience. AR technology has now developed to the point that it can be reliably used in real world applications. For example, medical data can be superimposed onto a patient's body to give a surgeon x-ray vision (State *et al.*, 1996), virtual planets can appear in an educational astronomy application (Shelton and Hedley, 2002), and games can be played in which artificial characters appear in the players real environment (Piekarski and Thomas, 2002).

As AR technology has matured it has also been applied into the architectural domain. One of the most obvious applications is visualisation of buildings on site. One of the first examples of this was in 1999 with the work Höllerer *et al.* (1999) in which a person could walk around a real university campus and see virtual models of buildings that used to be at that location many years ago. Since that time similar work has been performed in the Archeoguide project to show archaeological models (Vlahakis, 2002) and by Thomas *et al.* (1999)

X. Wang and M.A. Schnabel (eds.), Mixed Reality in Architecture, Design and Construction, 93–104.

to view buildings about to be constructed amongst others. AR visualisation can also be used to provide greater understanding of existing buildings. Feiner demonstrated an AR interface that reveals hidden architectural details indoors, such as framing details behind walls (Feiner *et al.*, 1995). More recently Dunston *et al.* (2002) use AR technology for visualising AEC designs. A number of others have also produced interesting demonstrations of AR technology applied to architectural visualisation.

A second application area is in aiding the design and construction process. An obvious extension of onsite visualisation is being able to design onsite as well. This was achieved when Piekarski (2004) developed a wearable AR system that allowed him to use AR computer aided design outdoors. In this way it was easy for him to be able to add virtual extensions to real buildings and walk around them to examine them from any perspective. AR technology can also be used to assist with the architectural construction process. For example, Webster *et al.* (1996) developed an AR assembly application that used vir tual cues to show step by step how a real space frame structure should be assembled. Others have also conducted research in AR architectural design and construction tools.

Over the next few years new technology will create further developments in these areas, as well as enabling other types of AR architectural applications to be explored. One of the biggest trends will be in moving AR experiences from fixed desktop computers to mobile devices, and in particular mobile phones. In this chapter we describe recent research results in mobile phone based Augmented Reality and show how these may be used in the architectural domain. We first begin with a review of previous work in mobile AR, then describe how mobile phone based AR provides new opportunities for architectural applications, and give possible future research directions.

2. Mobile AR Technology

2.1. BACKGROUND

As described in the introduction, in the mid 1990s researchers began to explore the use of wearable computers for mobile AR applications. The MARS (Mobile Augmented Reality Systems) project (Feiner *et al.*, 1997) from Columbia University was one of the first mobile augmented reality systems which allowed the user to freely walk around while carrying all the necessary system hardware. Several other prototypes such as BARS (Julier *et al.*, 2000), Tinmith (Piekarski, 2004) and mobile Studierstube (Reitmayr and Schmalstieg, 2001) have also been used for mobile AR architectural applications. For example one application based on the Tinmith system allowed people to construct virtual buildings onsite outdoors.

However, these systems had the common disadvantages that they were bulky, could only be used for short periods of time due to limited battery life and often had novel user interfaces that were difficult to learn. The original Columbia University system weighed over 40 pounds and was built on a custom wearable PC, GPS hardware, inertial head tracking system and see-through head mounted display (Figure 1), while the Tinmith system used a custom glove based input device which took time to learn. It is difficult to imagine systems this complex being used for long periods of time, or being able to be mass produced and used by many people.

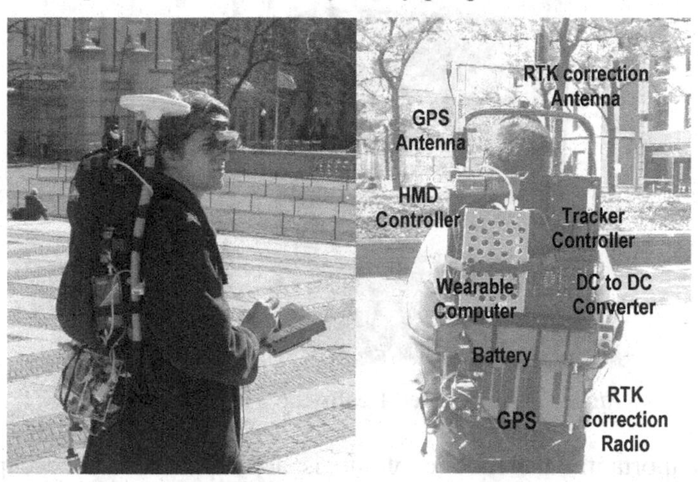

Figure 1. The Columbia University backpack AR system.

This reliance on bulky hardware decreased in 2003 when Wagner ported the popular AR tracking library ARToolKit (ARToolKit, 2008) to the Pocket PC platform and created the first self contained PDA AR application (Wagner and Barakonyi, 2003). Since that time the processors in mobile phones have become fast enough to also support AR applications. In 2004, Möhring (Möhring *et al.*, 2004) created the first mobile phone AR application while Henrysson and Ollila (2004) ported ARToolKit to the Symbian platform. To achieve this they created custom computer vision libraries that allowed developers to build video see through AR applications that run on a mobile phone. Therefore, since 2004 it has been possible to develop AR applications that run on consumer level mobile phones. Figure 2 shows a sample image of an AR application running on a Nokia mobile phone.

The mobile phone is an ideal AR platform because the current phones have full colour displays, integrated cameras, fast processors, and even dedicated 3D graphics chips. Wagner (2007) identifies the following advantages for PDA and mobile phone AR:

- The devices have low per-unit costs.
- A compact form factor.
- Low weight allowing comfortable single-handed use.
- The touch screen enables the creation of intuitive user interfaces.

- There is no need to use head mounted displays which are expensive fragile and difficult to handle.

Figure 2. Augmented Reality on a mobile phone.

Most importantly the mobile phone is a ubiquitous device capable of providing an AR experience to millions of users. In the past, the cost and complexity of AR hardware prevented widespread use of AR applications, but with a mobile phone people already have the required hardware in their pocket.

2.2. MOBILE AR APPLICATIONS

There are many possible AR applications that can be shown on a mobile phone. Henrysson and Ollila (2004) and Möhring *et al.* (2004) have shown how mobile phones can be used for simple single user AR applications. These first applications allowed users to load virtual models, find the phone camera position, create graphics and compose virtual images with the live phone video view (Figure 3).

These applications work in a similar way to desktop AR applications. The phone camera is used to stream video of the real world and image-processing software on the phone determines the pose of the camera (and thus the phone) relative to known markers or visually determined features. Once the camera position is known then a virtual object can be drawn with a virtual camera set at the same position and the graphics combined with the live video stream to create the illusion that the virtual object is part of the user's real world. The graphics are typically drawn using the OpenGL ES library (Khronos Group, 2008) a mobile variation of the popular OpenGL graphics library, or based on a higher level rendering library such as EdgeLib (Elements Interactive B.V., 2005/2008). Figure 4 shows the typical software architecture for a mobile AR application.

Figure 3. Model viewing using a mobile phone.

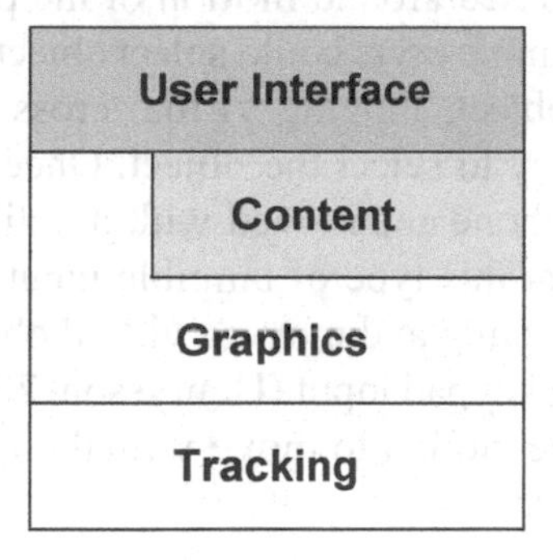

Figure 4. Mobile AR software architecture.

2.3. MOBILE AR USER INTERFACE

The main difference between mobile phone and desktop AR experiences is in the user interface. Instead of wearing a head mounted display, the user looks at the phone screen, and a keypad or stylus is typically used for input. In earlier work (Henrysson *et al.*, 2005a, 2005b) we identified the following key differences between using a mobile phone to a head mounted display (HMD) based system:

- The display is handheld rather than head worn.
- The phone affords a much greater peripheral view.
- With a phone, the display and input device are connected.

These differences mean that interface metaphors developed for HMD and desktop based systems may not be appropriate for mobile phone based AR systems. For example, desktop AR applications often assume that the user has both hands free to manipulate physical input devices which will not be the case with mobile phones. One metaphor that we have used is to assume the phone is like a handheld AR lens giving a small view into the AR scene. With this in mind we assume that the user will be more likely to move the phone-display to look at different parts of the real world rather than change their viewpoint relative to the phone.

For architectural applications it may seem a disadvantage to use a small screen to view AR content. However, recent studies by Hwang *et al.* (2006) have found that being able to freely move the screen around creates a sense of immersion equivalent to a fixed screen many times as large. In these experiments they compared the sense of presence felt when viewing an immersive virtual environment on a fixed 42 inch plasma screen, to a handheld 6 inch screen that could be freely moved. Users reported that they felt as immersed in the content shown on the handheld display as with the fixed screen. Thus we can assume that an AR experience viewed on a mobile phone display, may be as compelling as one viewed on a fixed large screen.

Assuming that the user is going to use the phone as a handheld AR lens, input techniques can be based around motion of the phone itself, rather than just keypad input. For example users could select objects by pointing the phone in the direction of the object, moving virtual cross hairs over the desired content, and pressing a key to select the object. Once an object is selected it could be attached to the phone and moved with it until it is dropped again. In user studies we found that this type of tangible input technique was a much more efficient way of selecting and moving virtual objects in a mobile phone AR system, than by using keypad input (Henrysson, 2005b). Subjects were up to 50% faster using phone motion to move virtual objects than using keypad input.

2.4. SAMPLE MOBILE AR ARCHITECTURAL APPLICATIONS

Mobile phone AR applications are relatively recent, but there have been several prototypes presented that show how the technology could be applied in an architectural and design setting. One obvious application area is for architectural visualisation. For example Mendez *et al.* (2006) have developed an application that allows people to look at a real architectural model with their phones and see a virtual representation of the pipes and underground utilities superimposed over the model. The user can see layers of information that is usually presented on different real pages shown virtually in the context of the real buildings. Figure 5 shows the user's view of the AR enhanced model. It is easy to imagine how a variation of this application could be used outdoors to allow a person with a mobile phone to see beneath the real sidewalk and look at virtual utility pipes below.

Figure 5. Mobile AR architectural visualisation.

One interesting variation of a mobile AR application for visualisation is Nokia's MARA project (Kähäri and Murphy, 2006/2007). In this case, Nokia researchers were exploring how a mobile phone could be used to replace bulky wearable computers and still provide a compelling outdoor AR experience. However, computer vision tracking alone is not sufficient to find the phone position outdoors. In this case they developed a hybrid tracking solution that added an inertial sensing pack to the phone. When combined with the phone GPS data this gives accurate position and heading data. The demonstration application shown with MARA is a mobile navigation tool, which overlays virtual cues on the real world to show the desired path. This technology supports a top down mode where the user's position and viewing angle is show overlaid on an aerial photograph of their immediate region. Figure 6 shows the MARA application in use. The same hybrid tracking technology could be used as the basis for onsite viewing of virtual buildings in place.

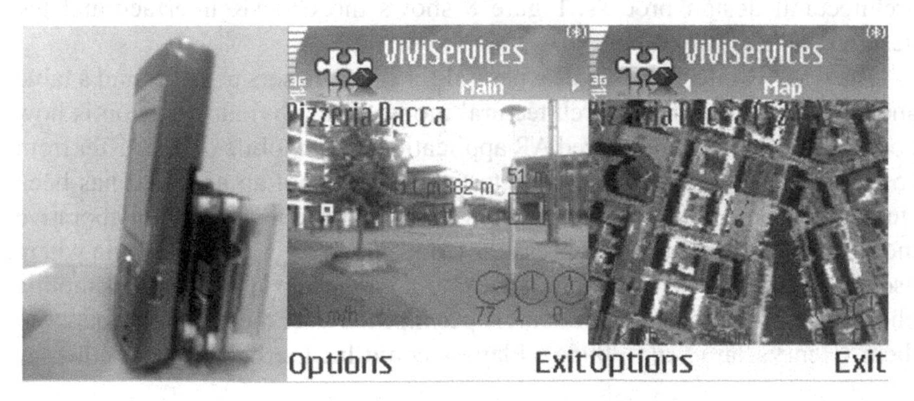

Figure 6. Nokia MARA Interface.

There have also been previous examples shown of how mobile phone AR technology could be used to support design. We developed a simple scene assembly program that allowed users to build AR scenes using blocks (Henrysson *et al.*, 2005a). This was designed to test how intuitive a tangible input metaphor was for AR interfaces on mobile phones. Users were able to select virtual blocks and move them around by moving the real phone. Figure 7 shows a typical sequence of moving and arranging virtual blocks.

Figure 7. Mobile AR block arranging.

A more complex mobile AR architectural design application is CMAR (Andel, 2006). In this phone application up to four users could collaborate together to arrange virtual furniture on the real printed floor plan of a house. Users could select the piece of furniture that they were interested in, then use a tangible interface technique to virtually attach it to their phone and drag it to the desired location. Bluetooth was used to wirelessly update the scene model and share it with all the other users so that they were working with a consistent model. Another unique element with CMAR was that the phone AR interfaces were used to arrange the virtual scenes, but the scene itself was also rendered on a PC for higher visual quality and shown on an external presentation monitor. This shows how a mobile AR interface could be used to support the architectural design process. Figure 8 shows the CMAR interface and the design results.

The CMAR interface makes it natural for several users to sit around a table and interact with a shared architectural model. One obvious question is how does collaborating using shared AR applications on a mobile phone differ from collaboration with a desktop AR interface. This is not an area that has been studied in depth, but we have conducted a small user-study with a collaborative mobile phone AR game. In this case, we developed an AR tennis game where users could sit across the table from one another and use their real mobile phones to view a virtual tennis court superimposed over the real world between them (Henrysson *et al.*, 2005c). Players could hit the ball to each other by

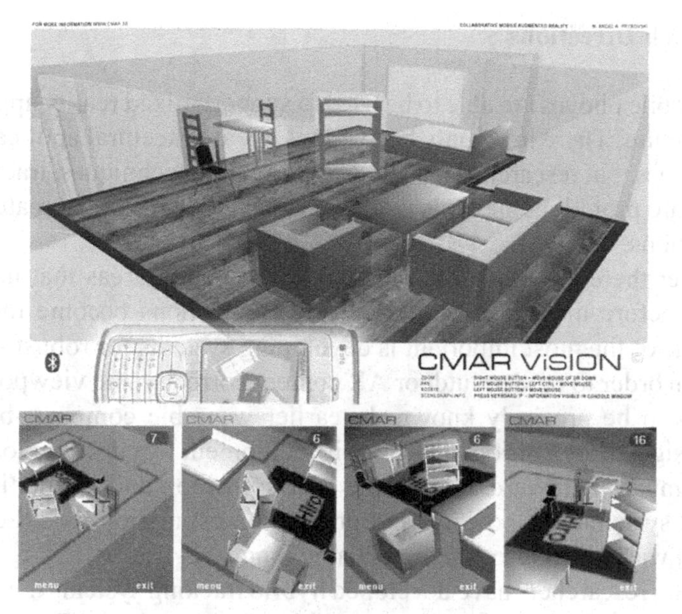

Figure 8. CMAR interface.

moving their phone in front of the virtual ball (Figure 9). We compared people playing in three conditions: (1) in an AR mode, (2) in a graphics only mode where the user did not see video of the real world on their screen, and also (3) in a non face-to-face condition. Users overwhelmingly preferred the AR condition because they felt that they could more easily be aware of what the other player was doing and collaborate with them. Although not an architectural application, the results from this experiment imply that similar benefits may occur in non-gaming shared mobile AR applications.

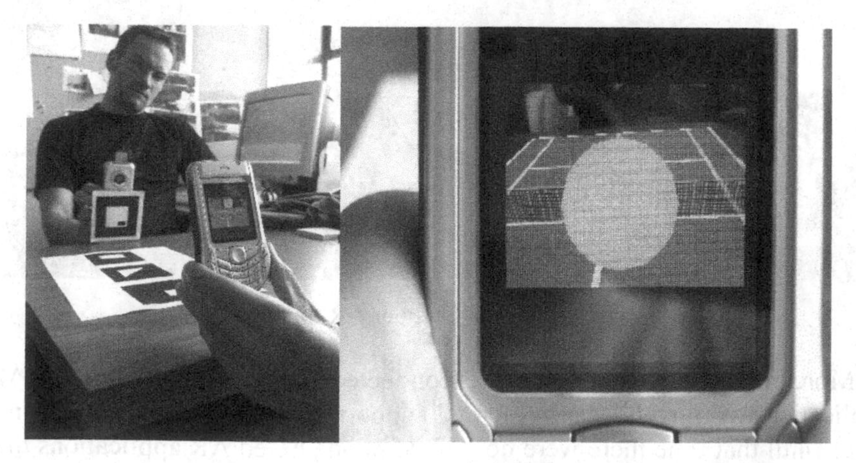

Figure 9. AR tennis interface.

3. Research Directions

Current mobile phones are able to be used to support mixed reality applications for architecture. There have only been a few AR architectural applications for mobile phones, but research in mobile AR interaction techniques, tracking, and rendering are providing the tools that can be used to develop a greater variety of applications.

However there are a number of important research areas that need to be addressed before mobile phone based AR applications become more wide spread. One of the most important is developing systems for robust wide area tracking. In order to render outdoor AR content correctly, the viewpoint of the user needs to be precisely known. In earlier wearable computer-based AR systems a significant portion of the mobile equipment was devoted to GPS and inertial compass hardware to track the user's position and orientation. Even with these systems, the position and orientation estimation would drift over time and it was difficult to support reliable long term use.

Recently, researchers have developed hybrid tracking systems that combine input from several sensor systems to provide robust outdoor AR tracking. For example, Reitmayr and Drummond's (2006) research combines GPS with inertial compass and computer vision input on a handheld device. By using a coarse virtual model of the real environment they can take the user's GPS location and compass heading and calculate what the user should be seeing. They can then match an artificial view of the environment generated from that position with the view from the real camera to provide error correction. The final outcome is robust tracking over a wide area, allowing them to overlay virtual content on real buildings (Figure 10). This is just the beginning, and it is clear that more research needs to be performed in this area.

Figure 10. Robust outdoor AR tracking.

More research also needs to be conducted on scalability. The CMAR application was novel because it could support up to eight users at the same time. Until that time there were no mobile phone based AR applications that had demonstrated how to scale to more than four users. However, once the technology becomes available to potentially hundreds of thousands or millions of users then there will need to be significant research conducted on scalability to support the data sharing between the end-users. Luckily this same problem is being tackled by researchers developing massively multiplayer games for mobile phones.

Finally, research should be conducted on the types of architectural applications that are possible when it is possible to track many users. The locations of mobile phones are known to the network service provider through the cellular towers that they are connected to. So if a large proportion of the population in a city have mobile phones which are turned on then the service provider can track population movements. This is exactly what Calabrese and Ratti have done in the Real Time Rome project from the MIT SENSEable City Laboratory (Calabrese and Ratti, 2007). In this case they collaborated with Telecom Italia to get access to near real-time information about cell phone location in the city of Rome. They then used this to visualise population density and flow on an hourly basis (Figure 11).

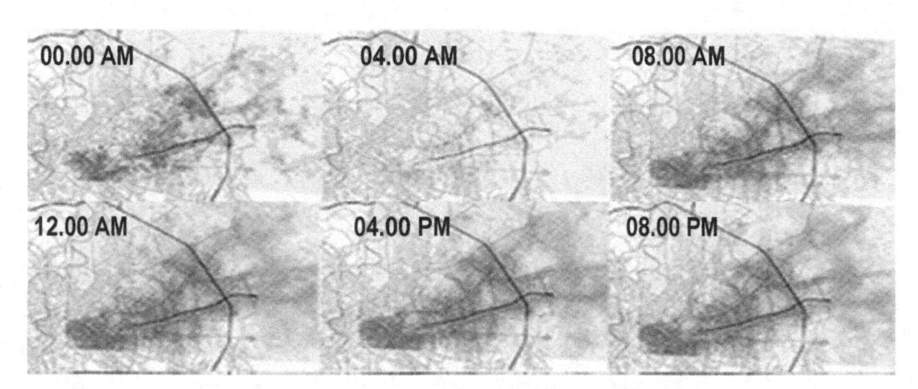

Figure 11. Real Time Rome population density.

Until now we have described mobile phone AR interfaces that can be used to provide an AR experience for the end user. However tracking and visualising movements of large numbers of phones could provide an architectural mixed reality experience on an urban scale. The individual people in the city in effect become movement sensors that can be used to monitor traffic flow and population density. This could be used by city planners and urban designers to create city scale mixed reality applications. However, this work is still at a very early stage and more research will need to be conducted before its potential is fully realised.

4. Conclusions

In this chapter we have described how mobile phones can be used to provide an architectural AR experience. As can be seen, mobile phone technology has developed to the point where it can replace the bulky wearable computers used for outdoor AR applications ten years ago. Computer vision tracking technologies, graphics libraries and interface tools have been ported over to the mobile phone making it possible to create applications that until recently could only be run on desktop machines. Since this foundational technology is now in place there is an opportunity for the architecture and design communities to create a wide variety of AR applications, especially with continued advances in hybrid tracking technologies.

Unlike earlier systems the widespread use of mobile phones means that there is also an opportunity to explore large scale architectural AR applications. The work of Calabrese and Ratti (2006) is one example of this. For the first time, urban designers have the ability to track entire city populations in near real time. Overlaying this data back onto the real city could provide some unique mixed reality architectural applications. The use of mobile phone based AR technology will enable users to experience architectural mixed reality applications in their everyday lives, both on an individual and community level.

AUGMENTED REALITY VISUALISATION FACILITATING THE ARCHITECTURAL PROCESS

Using Outdoor Augmented Reality in Architectural Designing

BRUCE HUNTER THOMAS
University of South Australia, Australia

Abstract. This chapter presents an overview of how augmented reality can improve the visualisation of architectural designs. An overview of wearable computer technologies and augmented reality is provided for a better understanding of the technology. The Wearable Computer Laboratory's Tinmith wearable outdoor augmented reality backpack system is described to reveal the current state of the art in this form of technology. The key contribution of this chapter is an explanation of how the user of a wearable augmented reality computer system can facilitate the architectural design process.

Keywords. Augmented Reality, Wearable Computers, Architecture, Visualisation.

1. Introduction

The revolution of wearable computers (Bass *et al.*, 1997; Thorp, 1998) and light-weight head mounted displays (HMDs) over the past ten years has made it practical to take augmented reality (AR) (Azuma, 1997; Azuma *et al.*, 2001) into the outdoors (Feiner *et al.*, 1997; Piekarski and Thomas, 2003c). AR is the process of a user viewing the physical world and virtual information simultaneously, whereby the virtual information is registered to the physical worldview. AR has been employed in a number of domains, such as the military (Julier *et al.*, 2000), surgery (Fuchs *et al.*, 1998) and maintenance work (Curtis *et al.*, 1998; Feiner *et al.*, 1993). Bringing AR outdoors requires the coupling of global positioning system (GPS) receivers and digital orientation sensors with 3D graphical models. Systems such as Tinmith (Piekarski and Thomas, 2001) are spatially aware computer systems for mobile users working

X. Wang and M.A. Schnabel (eds.), Mixed Reality in Architecture, Design and Construction, 105–118.

outdoors. I anticipate outdoor users requiring hands-free operation, and related AR applications are therefore particularly well supported by wearable computers and non-traditional input devices. A motivating application of these mobile AR systems is the visualisation of new architectural designs at the actual building site. This chapter discusses the process of using mobile augmented reality to improve the understanding of architectural designs through improved visualisation. Several systems are discussed, but the focus of the chapter is how Augmented Reality may help with this process. A number of constructed concept-demonstrator systems along with conversations with architects have determined the processes, which have been distilled down to a set of key points.

1.1. AUGMENTED REALITY

The key to making in-situ visualisation of architectural designs practical is augmented reality technology. Figure 1 depicts how AR works: the user's normal visual stimulus of the physical world is combined with computer-generated images. An optical combiner via video camera images fused with graphical images by the graphics chip set on a notebook computer was employed. The final fused image is presented to the user through a traditional VR HMD. Figure 2 depicts an example of an AR view from the Tinmith system. Unlike VR, where the computer generates the entire user environment, AR places the computer in a relatively unobtrusive assistance role. Using a wearable computer with a video see-through HMD allows people to move freely while working. Using GPS and orientation sensor technology the computer gains an additional and important input, the user's location, and thus computer applications gain spatial awareness that remain synchronised with the user's own awareness.

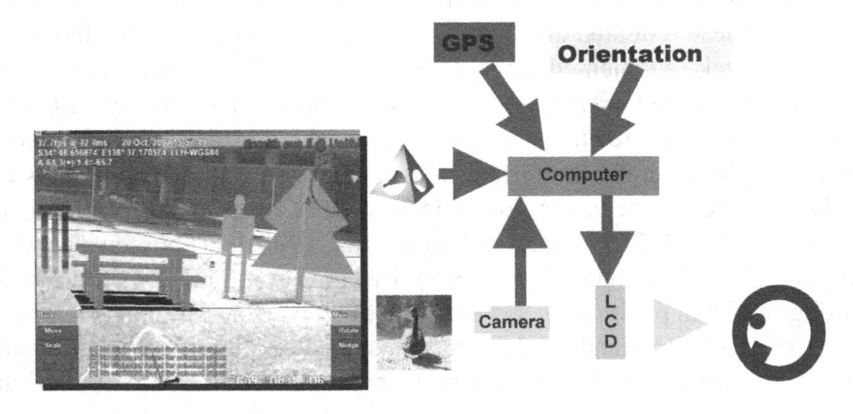

Figure 1. Overview of Augmented Reality.

Figure 2. An example view from Tinmith.

1.2. THE PROBLEM

How does one visualise the architectural design for a new building or a modification to an existing building relative to its physical surroundings? In the past, technical plans would have been made and models built. Current use of CAD packages extended this process to visualise the design of the building fully rendered as a 3D graphical model on a traditional workstation. Changes may be completed whilst the customer is in the design studio and the result may be visualised for the duration of this process. With the advent of Virtual Reality (VR), visualisations that are more ambitious were made possible. VR enables customers and designers to view a design in an immerse environment (Brooks, 1986; Mine and Weber, 1995) with the use of a VR head-mounted display. People are placed in a simulation and then simulate walks through the new design. They can visualise and move through the layout of the building in 3D. Tracking each user's head, allows for an intuitive movement of the head to change the viewing direction. Treadmills allow users to move by walking through a design while still physically inside the design studio. Together, tracking and treadmills allow users to sense the size and position of features in a new design. However, how can a user place a new building or extension in context with the proposed building site? Digitally enhanced photographs can show the placement of a building with respect to one vantage point. Models may be built to provide more vantage points, but these are expensive and time consuming to create, and offer only an artificial rendition of the site.

1.3. THE SOLUTION

The solution that the Wearable Computer Laboratory employs is to allow a user to walk around the site where the new building is to be constructed and visualise this new artefact in the spatial context of the existing environment. AR may be employed as a technique to provide this visualisation. AR has been used before in visualising interior design information. Webster *et al.* (Webster *et al.*, 1996) developed AR systems to improve methods for the construction, inspection, and renovation of architectural structures. Their initial experimental AR system shows the location of columns behind a finished wall, the location of re-bars inside one of the columns, and a structural analysis of the column. Like other researchers, Azuma *et al.* (1999) and Feiner *et al.* (1997), the Wearable Computer Lab of the University of South Australia is taking this use of AR from the indoor setting and placing it in the outdoor environment. The Tinmith system has been employed as a mobile AR platform to display architectural designs in an outdoor environment (Piekarski and Thomas, 2001; Thomas *et al.*, 1999).

2. Background

Wearable computers have now progressed to the processing power available on desktop computers. Such systems are commercially available and combined with an HMD deployed to assist workers with tasks that require information to be presented while keeping the hands free. Systems have been tested in the field with studies such as those by Siegel and Bauer (Siegel and Bauer, 1997) and Curtis *et al.* (1998).

A key feature of a wearable computer is the ability for a user to operate the computer while being mobile and free to move about the environment. When mobile, traditional desktop input devices such as keyboards and mice cannot be used, and so new user interfaces are required. Thomas *et al.* performed a survey of various input devices for wearable computers and how they could be used for collaboration tasks (Thomas *et al.*, 1998). Some currently available devices include chord-based keyboards, forearm-mounted keyboards, track-ball and touch-pad mouse devices, gyroscopic and joystick-based mouse devices, gesture detection of hand motions, vision tracking of hands or other features, and voice recognition.

The first demonstration of AR operating in an outdoor environment was the Touring Machine by Feiner *et al.* (1997) from Columbia University. The system is based on a large backpack computer system with all the equipment necessary to support AR attached. The Touring Machine provides users with labels that float over buildings, indicating the location of various buildings and features at the Columbia campus. Interaction with the system is via a GPS and head compass to control the view of the world, and when gazing at objects of interest longer than a set dwell-time the system presents further information. Further interaction with the system is provided by a tablet computer with a web-based browser interface to provide extra information. The Touring Machine was then extended by Höllerer *et al.* (1999) for the placement of what they termed Situated Documentaries. This system is able to show 3D building models overlaying the physical world, giving users the ability to see buildings that no longer exist on the Columbia University campus.

The Naval Research Laboratory is investigating outdoor AR with a system referred to as the Battlefield Augmented Reality System (BARS), a descendent of the previously described Touring Machine. Julier *et al.* (2000) describe the BARS system and how it is planned for use by soldiers in combat environments. In these environments, there are large quantities of information available (such as goals, waypoints, and enemy locations) but presenting all of this to the soldier could become overwhelming and confusing. Using information filters, Julier *et al.* demonstrated examples where only information of specific relevance to the user at the time is shown. This filtering is based on the user's current goals, and their current position and orientation in the physical world. The BARS system has also been extended to perform some simple outdoor modelling work (Baillot *et al.*, 2001). For the user interface, a gyroscopic mouse is used to manipulate a 2D cursor and to interact with standard 2D desktop widgets.

Apart from the previously mentioned systems, a small number of other mobile AR systems have also been developed. Billinghurst *et al.* (1998; 1999) performed studies on the use of wearable computers for mobile collaboration tasks. Yang *et al.* (1999) developed an AR tourist assistant with a multimodal interface using speech and gesture inputs. The TOWNWEAR system by Satoh *et al.* (2001) demonstrated high precision AR registration using a fibre optic gyroscope.

3. The Tinmith System

The Tinmith system is an outdoor augmented reality wearable computer system, and we have produced a number of demonstration applications (Piekarski and Thomas, 2003a, b). These applications use a glove-based menu system, image-plane manipulation techniques, and a new model creation methodology called 'construction at a distance.' This novel method allows users to construct 3D models of remote objects by walking around the object, but without actually touching it or being close to it – the only requirement is that it is visible. The Tinmith system forms the base on which we wish to investigate mobile through-walls collaboration systems.

The Tinmith backpack, as of 2006, is lighter and more robust than our previous systems. We have taken our eight years of experience in the field and built a system using the best components that are currently available and have designed our own custom housing to make the system robust for use in outdoor conditions. The whole system weighs 4 kg. Battery packs are an additional weight of approximately 2–4 kg depending on operating time and battery technology used. The profile of the system is almost to the point where a large jacket can be worn over the top that would conceal the system. The photos in Figure 3 show the left side (with ventilation fan and power switch), and the right side (with antennas and helmet connector).

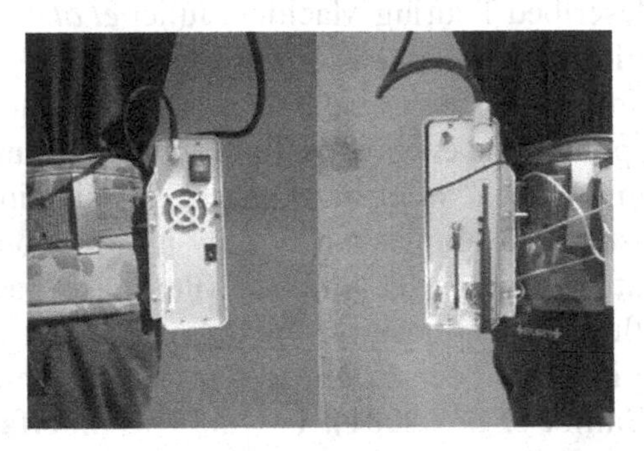

Figure 3. New Tinmith backpack.

The images in Figure 4 show the complete system from the front and rear with the all components visible. Note the lack of complex cables and the compact size of the unit. The batteries on the front are hot-swappable during operation, and two 8,000 mAh Ni-MH batteries are used for 3 h of operation.

Figure 4. Full system.

The entire system can automatically start doing AR visualisation within 70 s of turning on the power switch. Figure 5 depicts the Tinmith Gloves that are the main input devices for the user interface. This is composed of a set of pinch gloves for menu selection and thumb tracking to control two different cursors (one for each thumb).

Figure 5. The Tinmith gloves.

The system contains a custom modified Pentium-M 2.0 GHz computer with Nvidia GeForce 6600 graphics. The processor is designed for mobile applications, and the graphics processor is capable of handling any complex rendering task with ease. We use sub-50 cm accurate GPS receivers to provide excellent position tracking outdoors, and an InterSense 3 cube for orientation sensing. The system also implements 802.11 for wireless networking, Bluetooth for wireless peripherals, wireless video output, and USB and VGA ports for debugging.

The Wearable Computer Lab of the University of South Australia has shown that augmented reality can be used to visualise architectural designs in an outdoor environment (Thomas *et al.*, 1999). A design of an extension of the Physics building on the Mawson Lakes campus of the University of South Australia was effectively represented with the Tinmith mobile augmented reality platform. As an illustration of gaining the sense of a design, the straightforward extension was designed with a height of only 3 m. Informal testing showed this design flaw immediately. Even with simple line drawings for the building from the 1999 system (see Figure 6), a general feeling of shape and size was portrayed to the user.

The current Tinmith system provides full 3D rendering of architectural designs. The Wearable Computer Laboratory contracted a local graphic artist to build a small town to allow us to experiment with laying out a large collection of buildings and houses. Although these models are not from a traditional CAD system, they allow us to investigate the ability to visualise a small town.

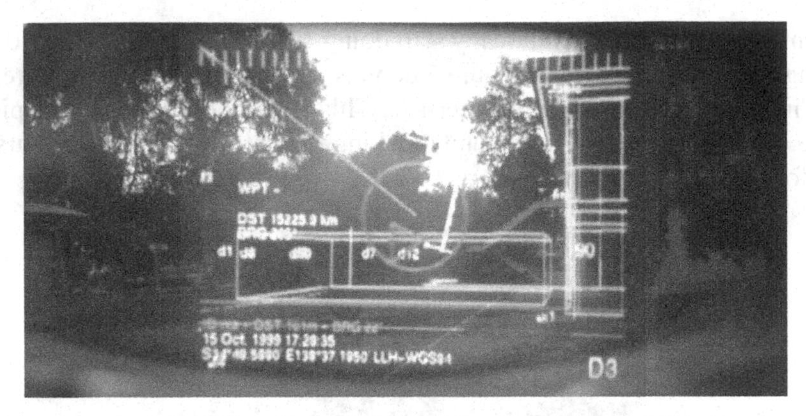

Figure 6. Tinmith architecture visualisation circa 1999.

Figure 7 depicts a house and car. A user is able to walk up to and around the house to gain a feeling of the size and shape. Figure 8 shows the ability to display finer detail as in the veranda on the side of a house. Figure 9 demonstrates the ability to visualise a street scene. The cars on the street are animated and are programmed to drive in a looping pattern along the streets. The scaling of the building appears to be incorrect in relation to the physical building. This is not the case, but it demonstrates a limitation of the system: correct occlusion must be performed. The physical building shown in the image was not defined in the town model, and the virtual buildings are in fact behind this physical building. Because the virtual buildings are much further away, the proper perspective scale shows them much smaller. The current Tinmith system renders all graphical images on top of the video stream. If the system does not

Figure 7. Town model of a house and car.

Figure 8. Town model of a veranda.

Figure 9. Town model of street.

know to occlude a graphical object, they will always be drawn on top of whatever physical object is being viewed. This can be avoided if models of the physical objects are incorporated into the graphical scene.

The Wearable Computer Laboratory has been investigating through-walls collaboration techniques. In particular they have developed a new interaction metaphor they have termed god-like interaction, or *Hand of God* (Stafford *et al.*, 2006). This metaphor was developed to improve collaboration between a user located outdoors using mobile augmented reality systems, and a user located indoor working on tabletop projected displays. The metaphor leverages an indoors user's ability to manipulate physical props as well as their hands onto a table surface that is then scanned as a 3D object and transmitted wirelessly to a remote user outdoors. Small objects such as models of buildings and houses can be placed on a map, Figure 10, and will be converted into 3D textured models. These 3D models are then sent as geo-referenced to the outside user who is then able to view them in the specific orientation and position in the physical world, Figure 11.

Figure 10. Operation of the Hand of God.

Figure 11. View from the HMD of the outside user.

4. Using a Mobile Augmented Reality Platform

The Wearable Computer Laboratory is investigating and developing computer technology that literally takes computers out into the field, where computer applications are geographically aware and designed to interact with users in their world, not just in the confines of the computer's artificial reality (Piekarski and Thomas, 2003a).

By providing information in a 3D form, in scale with the surroundings, AR systems provide significant benefits:

- Physical objects with known locations can be found more rapidly, especially in featureless terrain, thus saving time and costs. Imagine you are at a large building site with many services level to the ground, and you wish to locate a number of them. AR visual cues can be provided to make these objects easier to locate.
- In the case of an object being underground, the location of this object can be determined within the accuracy of GPS and orientation sensor systems. Locating underground pipes is a good example of how this could employed.
- Previously invisible features, such as boundaries, become visible without the use of physical markers. In the case of a housing development, this would help with locating and understand the different lots.
- Overlaying more than one information source allows the relationship between objects to be determined easily. In the above housing development example, information such as owners, prices, proposed house structures, and tax information could display in such a way as to make it easy to understand the relationship between the different lots. Being able to perform this assessment in-situ with the physical surroundings would help users with particular decision-making processes. Potential buyers would be able to make decisions that are more informed.
- Features can be viewed from orientations that are more appropriate to the task than a map or drawing may allow. Maps and drawings are inherently 2D while the physical world is 3D. Viewing a new house from the first person perspective in-situ provides additional information and context, and makes issues such as size, shape, and colour easier to understand.

5. The Role of a Mobile Architectural Visualisation System in the Design Process

As an illustration, a number of architects and I discussed the possibilities of using outdoor augmented reality for the visualisation of architectural designs. The Wearable Computer Lab had made a number of simple building models of extensions to one of the buildings on our campus. One of the extensions was a large two-storey room off the end of a lecture theatre. The building model was developed in AutoCAD, and the existing building was based on drawings obtained from the University of South Australia. I explained how a fire escape was not on the drawings. Our extension was to be built right on top of this fire escape staircase. This became pointedly obvious the first time we viewed our model of the extension with the Tinmith system. I remarked to the architects, "Of course you guys would never make such a mistake." They all laughed and stated that these sorts of mistakes do happen, and being able to catch them early would be of great benefit.

As illustrated, a major benefit of an outdoor augmented reality system such as the Tinmith system is that it helps people visualise architectural designs in their physical outdoor context. The Tinmith system is designed to meet the following objectives for such visualisations:

- Architectural designs should originate from standard CAD packages and be stored in standard interchange file formats.
- Architectural designs will be displayed relative to their physical site placement.
- Modifications can be made at the building site.
- The user interface must be easy and intuitive to use.

Such outdoor augmented reality systems are required to be consistent with two contemporary architectural design methodologies. The first is that the system must be able to import data from standard architectural design software packages. The heart of the system is the facility to visualise or see characteristics of the architectural design in the field, imparting to the user a feeling of how the architectural artefact will fill or change the physical space. The targeted end users of such a system are architects, engineers, designers, and clients.

An outdoor augmented reality system could be employed at a number of key points in the design and construction process:

- *Scoping the project:* When an architectural project first starts, some initial ideas can be quickly examined in-situ to understand the direction of the building better. Physical walkthroughs may be performed in any large flat area such as a parking lot or playing field (Thomas and Piekarski, 2003).
- *Team collaboration:* In many cases, numerous people are involved in the decision making process at the site where a building will be constructed. Augmented reality allows for a common visualisation of the design or engineering concepts for all parties. In these cases, an HMD might not be the appropriate display technology, and the Wearable Computer Laboratory has been experimenting with a tripod-mounted augmented reality display. An experimental system for visualising GIS data has been built (King *et al.*, 2005).
- *Determine the proper placement of the building:* A key feature of augmented reality is the ability to quickly visualise and convey understanding of virtual information in the context of the physical world. A clear example of this is the placement of a potential structure on a building site, which enables the user to understand the issues of where this structure will be located, potential problems, alignment with other structures or features, and to form a better understanding of the size and shape of the structure. The placement of the virtual building

allows a better understanding of the current shape and gradient of the building surface giving an insight into how the building surface needs to be reshaped.

- *Visualising conceptual designs in-situ:* The relationship of a house or building to its surroundings is critical for the overall design. The ability to visualise these structures in-situ with other buildings, vegetation, and landscape can greatly improve the overall outcome of the design. Current use of digital enhanced imagery is limited. There is only one viewpoint per image, and these images poorly portray the overall combination of these in-situ features. Multiple designs may be presented to the customer while on site. This enables the end-user to make more informed design choices.
- *Making modifications on site:* As the visualisation is performed on a computing system, design modifications and annotations may be applied on the building site. Decisions can be recorded and tagged to the relevant portions of the electronic design. Changes to paint colour and building materials easily be reviewed and recorded. In the early stages of the design, primary exterior 3D designs can be presented with the expectation of quick modifications. For example, structural parameters such as building heights can be adjusted with some simple editing. One architect I am current working with performs early designs on Google SketchUp[1] as this system provides him with a method for quick end design for early discussions with clients. He stated that current CAD systems require too much effort for such quick designs. These designs can be exported to Google Earth[2] enabling the architect to place the design in the correct location quickly. The augmented reality system can access the Google Earth data and display this to the client.
- *Visualising construction and engineering data on site:* In addition to architectural design information, construction and engineering, data may also be viewed via the augmented reality system on site. This design data is in a similar format to that of architectural designs. During the construction phase, this data may be viewed via augmented reality for the following reasons: review of progress with client, location-based technical details for an engineer or contractor, and planning for the next phase of the operation. The case of a team of engineers and architects discussing critical issues, the ability to view the designs with the current state of construction will enable a clearer understanding. This would hold especially true for the visualisation of key structures to be built at major milestones.

6. Conclusion

In conclusion, this chapter presented an overview of how augmented reality can improve the visualisation of architectural designs. An overview of wearable computer technologies and augmented reality was provided for a better understanding of the technology. The Wearable Computer Laboratory's Tinmith

[1] Google SketchUp http://sketchup.google.com
[2] Google Earth http://earth.google.com

wearable outdoor augmented reality backpack system was described to demonstrate the current state of the art in this form of technology and the following six key points in the design and construction process were detailed as areas in which this technology might be applied:

- Scoping the project
- Team collaboration
- Determine proper placement of the building
- Visualising conceptual designs in-situ
- Making modifications on site, and
- Visualising construction and engineering data on site

The key contribution of the chapter is a description of how the user of a wearable computer augmented reality system can facilitate the architectural design process in the areas elaborated showing the vast potential awaiting application.

Acknowledgements

I wish to acknowledge the members of the Wearable Computer Laboratory for creating and developing the Tinmith system. Dr. Wayne Piekarski led the Tinmith system project, and other key contributors are as follows: Ben Avery, Ross Smith, Ben Close, and Gary King.

SIMULATION OF AN HISTORICAL BUILDING USING A TABLET MIXED REALITY SYSTEM

ATSUKO KAGA
Osaka University, Japan

Abstract. Many historical timber and timber-framed buildings remain standing in Japan today. One way to promote an understanding of the characteristics of these buildings is through digital content produced using visual information related to the construction techniques and structure. The study of important historical buildings requires knowledge of various fields: architecture, history, sociology, culture, and technology. However, available data in these areas is inadequate for research. A digital archive, which could facilitate multiple uses, was considered for this study. One form of that such an archive could take is via Virtual Reality (VR), making an interactive simulation possible. The internal structure and system are regarded interactively. Each user can build, as a member of a group, one component if VR contents are used. Furthermore, if Mixed Reality (MR) technology of piling up on-the-spot photographic images and VR images on-site and in real-time were used, VR contents would be accessible and visible for people who are present on-site. Then, the historical building's interior (not normally visible), and information about its components, can be mastered in the usual manner – that is, synthetically on-site or off-site. Furthermore, this research addresses MR technology, which layers on-the-spot photographic images and VR images in real-time to realise a simulation for educational purposes related to that is related to an historical building on-site. The proposed 'Tablet MR,' construction of an experimental model, evaluation of accuracy, and suggestion of a system application possibility were performed.

Keywords. Historical Building, Virtual Reality, Digital Information, Mixed Reality, Tablet PC.

1. Introduction

Many historical wooden framework buildings exist throughout Japan. To enable anyone to understand the characteristics of these buildings, it is necessary to provide digital contents using visual information about their structures

X. Wang and M.A. Schnabel (eds.), Mixed Reality in Architecture, Design and Construction, 119–134.

and construction techniques. Such substantial digital content is useful in construction training, and can aid understandings of historical buildings through exhibitions. Previously, to introduce construction methods used in ancient Buddhist architecture in Japan, 3D physical models played an important role.

2. Building an Historical Archive Using Digital Information

Several digital archives have been made about historical buildings with wooden framework architectures (Tang *et al.*, 2001, 2002; Shih *et al.*, 2004). Timber frame construction is complex and information gathered and presented needs to incorporate detailed elements such as construction joints as well as larger scale information such as internal and external views. In this chapter, the contents and presentation methods surveyed are based on an example of a digital archive of an historical building with wooden framework architecture (Yeo *et al.*, 2003).

The first digital archive introduced is that created in Suzakumon in 1987. In the eighth century, Suzakumon existed as the main gate of Nara, an historical capital of Japan. There it was, so people say, that foreign envoys were greeted and sent off, and where the Emperor held celebratory events. Although this gate had disappeared, it was restored in 1997 even though no data related to the structure of the gate had survived. A 3D model was created based on a proposal that estimated the appearance of the framework and the detailed components, which in turn was based on ruins and archaeological information by the Nara Cultural Property Research Institute. Rules about the combination and processing methods of components were extracted and were added to the definition of the model on the computer, thereby producing a digital archive.

In the digital archive of Toshodai-ji in 1997, an animation was created using Hi-Vision to explain the detailed structure of a 3D model of an auditorium, which was later used as teaching material for architecture students. In the digital archive of Ueno Kanei-ji in 1998, a joint and framework construction was created using CG animation in a 3D model; this was also used as teaching material for architecture students. An illustration of the traditional architecture of Japan using still images and Hi-Vision is shown in Figure 1.

To date, most digital material related to historical buildings has been produced for specific purposes and has appeared in video media formats such as VHS, Hi-Vision video, and digital video. The study of important historical buildings occurs in various fields, such as architecture, history, and sociology and they are of cultural and technical interest. However, the available data is inadequate for addressing the varied demands of such diverse fields. A more flexible form of digital archive that is applicable to a variety of uses and which can respond to various media was created for the five-story pagoda in Kyoto To-ji.

Figure 1. Traditional architecture of Japan presented using a still image and Hi-Vision.

3. Preparation of the Digital Archive

The digital archive exemplified is that relating to the five-story pagoda in To-ji, Kyoto City that was used for carrying out an interactive simulation. To begin, data relating to the pagoda was surveyed using past research about interactive digital archives (Yeo *et al*., 2003). The building was, and remains, a national treasure and was registered with the World Heritage Foundation in 1994. It is used as the symbol of Kyoto and its height, of 54.8 m makes it the tallest wooden five-story pagoda in Japan. The present pagoda, built in 1644, is the fifth generation and the foundations date from the end of the ninth century. It was built making full use of the best construction technology of the Heian era. This pagoda has withstood earthquakes and typhoons, which are common in Japan, yet it is difficult to acquire detailed information on its structure from reference books.

Information about the pagoda was collected in cooperation with the cultural property public assistance section in the Kyoto Prefecture board of education. Based on this information, a 3D model featuring detailed internal and external perspectives was prepared. The photographic data obtained at the information-gathering stage was used for 3D-model data texture mapping. This technology renders the created data highly realistic. Based on old references the detailed 3D form at the time of erection was recreated and presented using texture mapping. The precision of this digital archive is useful for educational purposes.

Digitising the data in an archive alleviates the problem of information degradation and loss; moreover, it can promote the reuse of data. A digital archive records quality digital images and maintains them in the form of multimedia, databases et cetera, and becomes an important historical, cultural, and scientific resource. The media of the digital archive taken up in the preceding chapters have thus far been restricted to animation and still images. Virtual reality (VR) technology and its systems have been developed as effective means of design simulation or presentation in the environmental design field (Lou, 2003; Kaga, 2005). Therefore, a specific examination of a real-time simulation as the medium of the new digital archive was undertaken. This was done because it enables free movement inside a 3D virtual space. Further, real-time simulation enables construction of an interactive knowledge database using 3D space, as shown in Figures 2 and 3. The resultant digital archive was used for sharing interactive knowledge and television broadcasts. VirtoolsDev of Virtools was used as the base software to recreate a local environment in 3D space through real-time simulation.

DirectX API, which realises real-time rendering, was used on a PC, and equipped with a Virtools player as a plug-in to a commonly used web browser. With VirtoolsDev, visual programming is achieved by connecting the program put together for every function of the series 'Behaviour Building Block' as an event-flow. Further, the series, which is not prepared beforehand, can be further developed using C++.

Figure 2. Five-storied pagoda: outside (left), and inside (right) (Yeo *et al.*, 2003).

Figure 3. Present figure of five-storied pagoda (left). A rendition of the five-storied pagoda when it was first constructed (right).

The following items describe the interface functions of the scene (Table 1).

Table 1. Interface functions of the scene.

Interface function	Details
Walkthrough	Enables free movement in space while observing the surroundings. It supports the ability to view objects in many directions. Both the interior and exterior of the building can be viewed separately. If the interior view is too narrow, the viewing angle can be adjusted (Figure 4).
Section & plane display	Cuts the building to show a section at any point, elevation or plane. The section is displayed in a perspective mode to easily determine the depth (Figure 5).
Show or hide main materials	Shows or hides materials classified into groups. Concealed objects or materials, such as those by an external wall, become visible (Figure 6, left).
Parts information display	Shows basic information related to various literature, for example what was used and for what purpose (Figure 6, right).

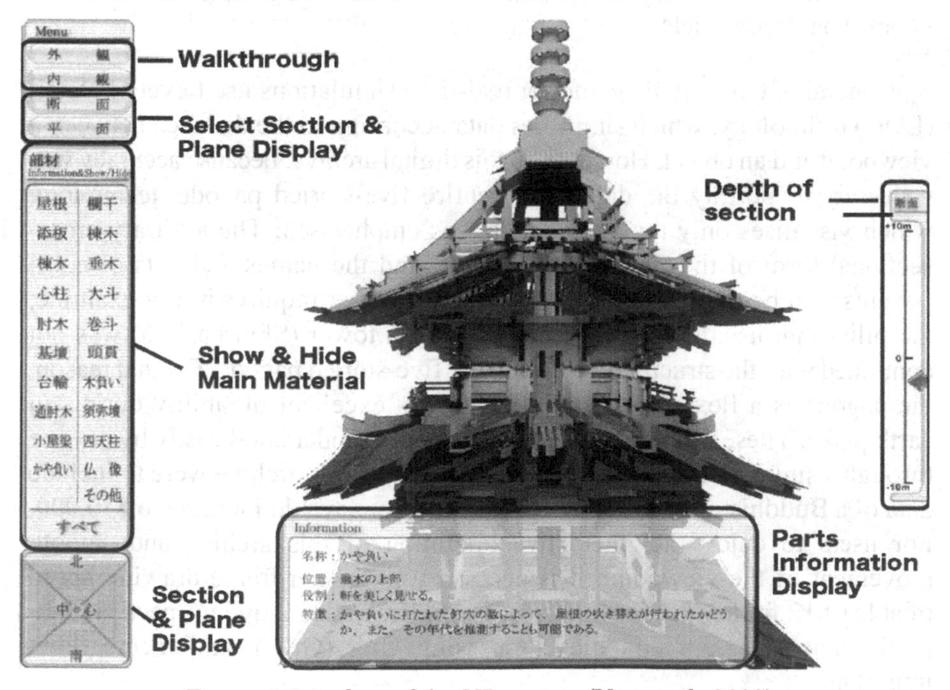

Figure 4. Interface of the VR system (Yeo *et al*., 2003).

Figure 5. Walkthrough outside (left), inside (Middle), section display, elevation (right) (Yeo *et al.*, 2003).

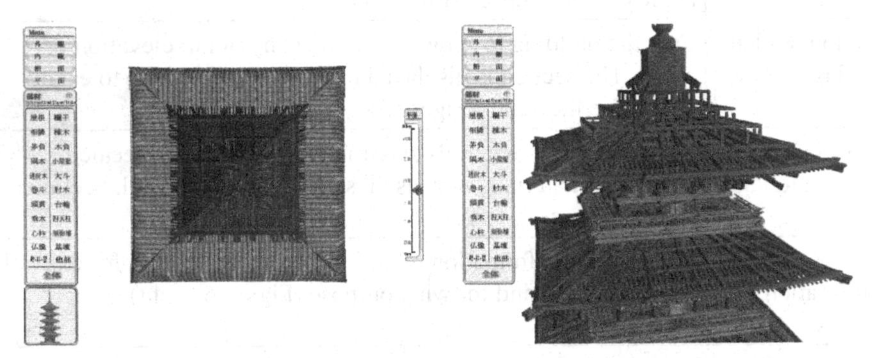

Figure 6. Section display, plan (left), show or hide main materials (middle) and Part information display (right).

Generally, existing 3D games or real-time simulations use Level of Detail (LOD) technology, which simplifies data according to the distance between a viewpoint and an object. However, in this digital archive, because accuracy was necessary to portray the data of the entire five-storied pagoda, technology which visualises only the visible area was emphasised. The arbitrary cross-sectional form of the five-storied pagoda, and the names and lists of components, can be displayed interactively when a user requires it. For example, the pillar that usually supports the centre of a tower (Shin-bashira) was not combined with the structural objects of the five-storied pagoda. For that reason, the pagoda is a flexible structure, which has excellent durability during an earthquake. These aspects of the structure can be understood easily by anyone through a simulation using a clipping function. If this archive were to include data of a Buddhist image, the number of polygons would increase to 830,000. For users to enjoy the interactive operation of this archive and smooth movement of the viewpoint, it is necessary to have a frame drawing speed of at least 12 frames per second (fps). To realise such a drawing speed, a high-performance CPU, a Graphics Processing Unit (GPU), and memory are important.

4. Simulation of Historical Buildings On-site

Real-time simulation of the historical building described in Section 3 is assumed for use at alternative sites, such as exhibition spaces. One problem with VR is that it is not possible to perform adjustment of a flexible human viewpoint, and understanding the scale or degree of slopes is difficult. When only ruins exist, and are used to build a 3D model, the historical building can be recreated in the real world intuitively by entering the physical space and uniting the VR image with on-the-spot photographic images.

When an historical building exists (because someone has visited the site and used VR and on-the-spot photograph images together), it is expected that the internal structure can be visually understood and a sense of its scale conveyed. Furthermore, if someone were to visit the site and use on-the-spot photographic images and the VR image united with Mixed Reality (MR) technology; it would be possible to add visual information to the framework construction or components. Because information can be understood on a scale of real space, it can be offered effectively.

4.1. MR TECHNOLOGY

MR is also called compound reality and Milgram and Kishino (1994) present an explanatory figure similar to that in Figure 7. Technical research related to MR has been carried out by Kato *et al.* (1999), You *et al.* (1999), and Behringer (1999) in relation to a technique that precisely unites a position – in VR with that the real site. However, few examples exist of effective use in the simulation of an historical building.

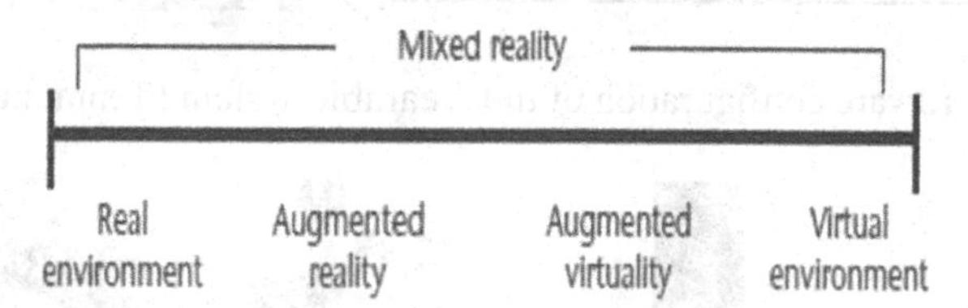

Figure 7. Relationship of reality from the virtual to the real (Milgram and Colquhoun, 1999).

4.2. PRIOR-ART SEARCH

The ‘Nara Palace Site Navigator’ is an example of a navigation system using Augmented Reality technology for an historical building using video see-through Head Mounted Displays (HMDs) and other devices. Research has also progressed toward the construction of a wearable sightseeing guide system of Nara Palace using multimedia contents, to which were added a restoration model, guide information, et cetera. N. Yokoya Laboratory in the Nara Institute of Science and Technology is advancing this research. The system distributes sightseeing guidance information in real time at a terminal with GPS, using not only an HMD but also a PDA and a cellular telephone. It consists of a small notebook PC and HMD, a gyroscope sensor, which detects a position and a posture, and a small video camera of the wearable computing type, as presented in Figure 8. Images of an old building are layered and displayed using Computer Graphics in three dimensions, as shown in Figure 9, on actual scenery, thereby providing a system with which a user can walk inside the building.

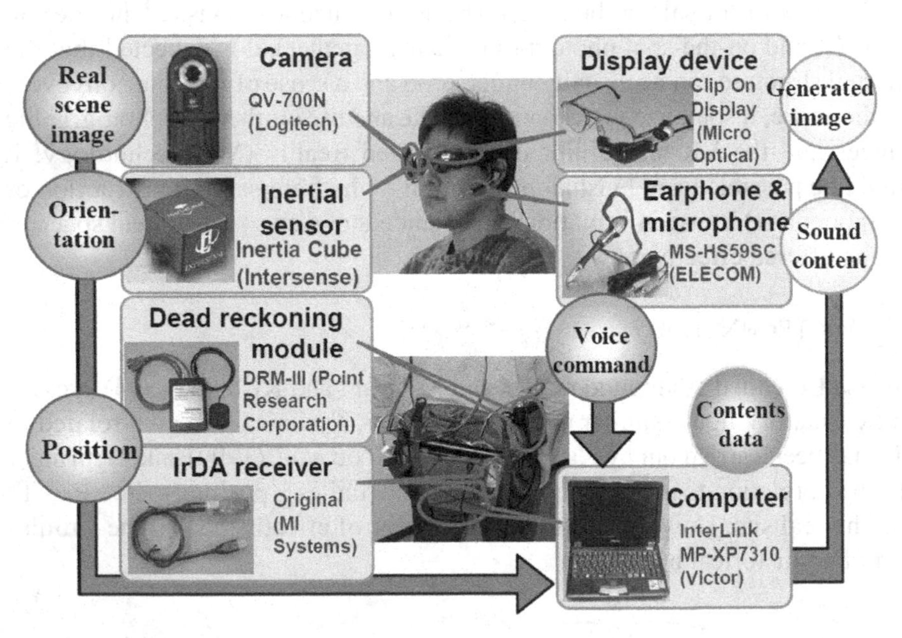

Figure 8. Hardware configuration of the wearable system (Tenmoku *et al.*, 2004).

Figure 9. Examples of user-presented images in the Nara palace site (Tenmoku *et al.*, 2005).

Figure 9a portrays the web map, which is generated automatically. Figure 9b,c show a presentation of intuitive sightseeing guidance information for the user, via notes attached to explain the actual environment using Augmented Reality. Further, in Figure 9c, image contents are depicted simultaneously with the synthetic picture. In Figure 9d, computer graphics are used to present an image of the building that existed in this place 1,300 years ago. It is compounded visually with the user's scenery via Augmented Reality technology.

4.3. RESEARCH AIMS

Ito (2005), described the availability of MR by examining a city space, and has classified MR according to characteristics like those shown in Figure 10.

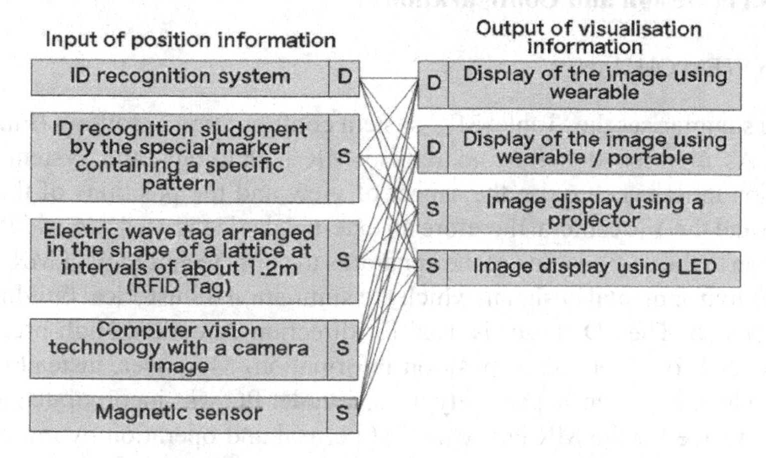

Figure 10. Input/output classification (D: Dynamic/S: Static) (Ito, 2005 – Appendix).

The DD type follows in the footsteps of operations in space, and mobile varieties, wearable, et cetera using an apparatus carried on the body for dynamic acquisition or presentation of information. Neither pattern matching nor an environmental embedding type sensor is used for this. Therefore, it is also considered to be applicable to object grounds, such as historical buildings.

To realise an on-site simulation, an experiment in system configuration was conducted and a prototype system was tested with particular emphasis on the MR technology of the grouping together of on-the-spot photographic images and VR images in real-time. This produced a simulation for with application to education related to an historical building on-site.

The proposed 'Tablet MR' involved constructing an experimental model, evaluating its accuracy, and making suggestions for its potential application. To realise MR technology, it is necessary to unify the positioning of a live camera, a VR camera, and their respective directions. For that reason, an external device was used. The apparatus for application of the MR technology in the field of an environmental design was considered: a 3D sensor, along with high-precision GPS and direction detection was used for the apparatus to detect position information. Tablet PCs are used more generally than HMDs; using the former, two or more people can peruse and operate the display of an MR image. Because each device is interlocked, the system was developed to enable consideration of data through external devices with VR software. A frame to fix a live camera, a 3D sensor, and a Tablet PC was designed and created, completing the MR system experimental model. In addition, the system was tested and verified, using an actual historic building.

5. System Design and Configuration

5.1. SYSTEM OUTLINE

Fukuda summarises the 'Tablet MR' system configuration as follows (Fukuda, 2006). As mentioned before, to realise MR technology, the system configuration must synchronise the aspect of view and the positions of the live camera and the VR camera, therefore an external device is used. Consideration was given to the composition of the apparatus to apply technology of MR in the field of environmental design in which the simulation of historical buildings is also included. The 3D sensor is used for direction detection. High-precision GPS is used for detection of position information. Moreover, instead of the HMD, which is currently generally used, a tablet PC was incorporated as the display device for the MR image so that perusal and operation by more than one person was possible. To interlock each device, software development was carried out to allow adjustment and use of each external device. An antenna and a receiver are required for GPS. Also needed are a PDA and a receiver for control of communication using a Bluetooth device. The interface for operating the MR space according to the purpose was implemented, Figure 11.

5.2. HARDWARE DESIGN

A frame incorporating a live camera, a 3D sensor, and a Tablet PC was designed and created, producing a "Tablet MR" system prototype, Figure 12. This was designed to be used with a tripod; therefore, it is operable even if a user must leave it to walk somewhere. The apparatus composition of the completed Tablet PC is shown in Figure 13.

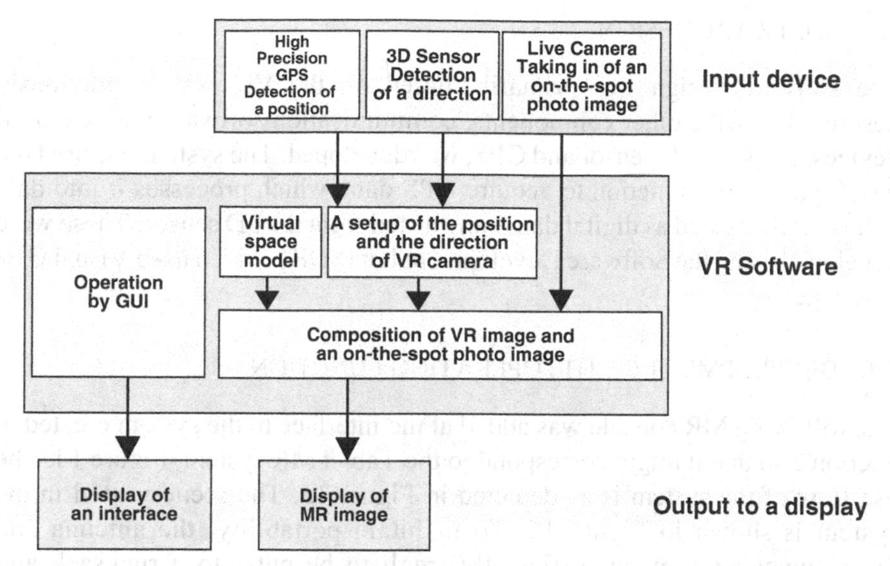

Figure 11. System outline (Fukuda, 2006 – Appendix).

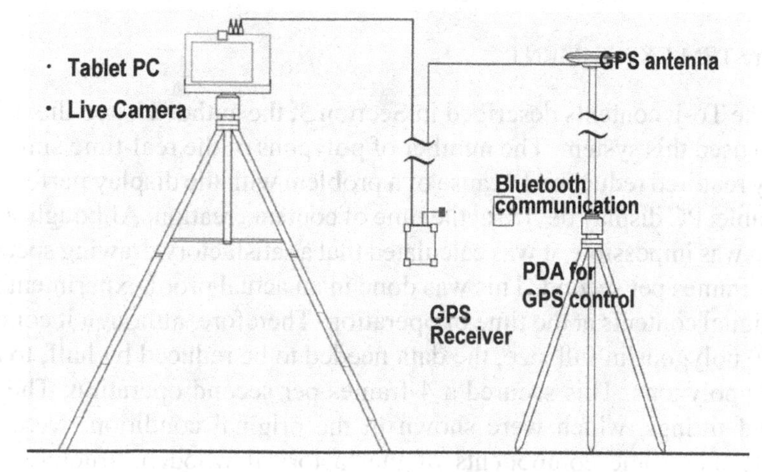

Figure 12. System configuration (Fukuda, 2006).

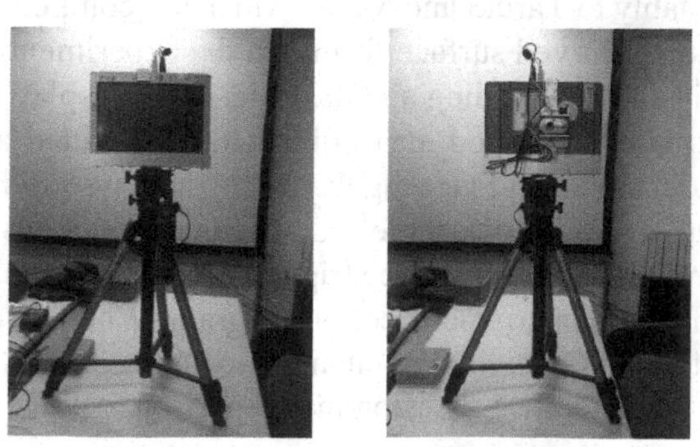

Figure 13. Tablet PC furnished with a live camera and a sensor.

5.3. SOFTWARE DESIGN

The software design was primarily based on the VR system previously described. For the other components, communication software with external devices, such as a 3D sensor and GPS, was developed. The system required the development of a function to acquire GPS data, which processes it into data that is easily treated as digital data transmitted from the 3D sensor. These were developed using the Software Development Kit (SDK), which used Visual C++ language.

5.4. DEVELOPMENT OF THE OPERATING FUNCTION

The following MR console was added at the interface to the system created in Section 3 so that it might correspond to the Tablet MR system, Figure 14. The use flow of the system is as depicted in Figure 15. The scenery used in the system is shown in Figure 16. To maintain portability, the antenna and the receiver were made sufficiently small to be put into a rucksack and carried. When the system is fixed to a tripod, it rotates.

5.5. SYSTEM EXPERIMENT

Using the To-ji contents described in Section 3, the author visited the original site and used this system. The number of polygons of the real-time simulation initially required reduction because of a problem with the display performance of the tablet PC display device at the time of content creation. Although smooth drawing was impossible, it was calculated that a satisfactory drawing speed was about 4 frames per second. This was done in an actual proof experiment using other digital contents at the time of operation. Therefore, although it contained 830,000 polygons in full size, the data needed to be reduced by half, to about 400,000 polygons. This secured a 4-frames per second operation. The outer wall and fittings, which were shown in the original condition, were made invisible. For some components of the historical wooden structure, transposition was made to a model which reduced the number of polygons. This occurred most notably to Taruki and Masu, which are components with many polygons due to their curved surface forms. In the experiment, the display of the components inside the building and the description display of components was carried out placing a tripod at a point which provided a general view of the five-storied pagoda, and moving the Tablet MR system in relation to it, Figure 17. The tap of a stylus on a tablet PC attachment operates the Tablet MR system. Therefore, the button-interface triggered a fundamental function such as GPS, a 3D sensor, or an error-correction program. When, in an experiment, the view angle of the VR camera was changed to 36 degrees, it was found that there was almost no difference in vision in comparison to a live camera image,

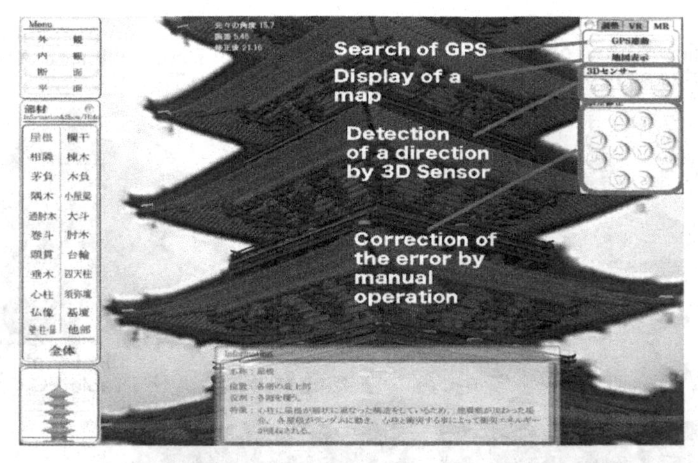

Figure 14. Interface of the 'Tablet MR' system.

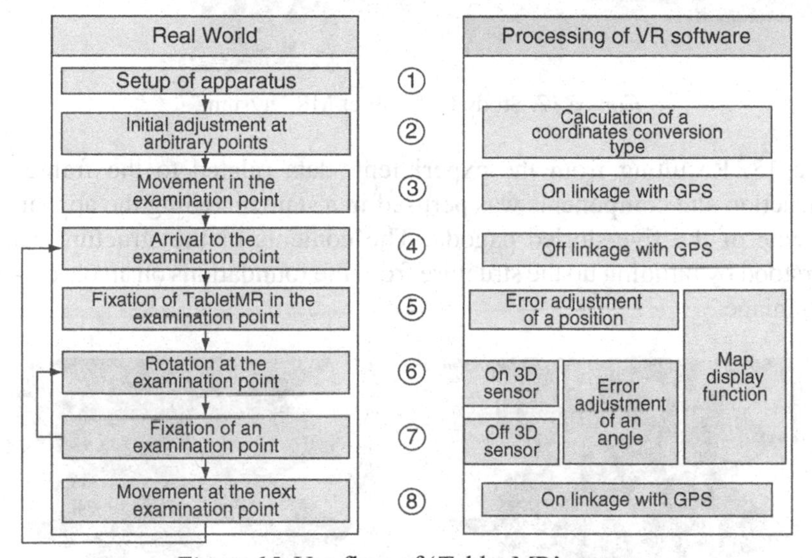

Figure 15. Use flow of 'Tablet MR' system.

Figure 16. Use scenery of the system.

Figure 17. Study by "Tablet MR" system.

Figure 18. Resulting from the experiment, data related to the framework construction and components was perused in a state matching the appearance and scale of the five-storied pagoda. The contents of the structure can be understood by building up the structure from the foundations on an on-the-spot photo image.

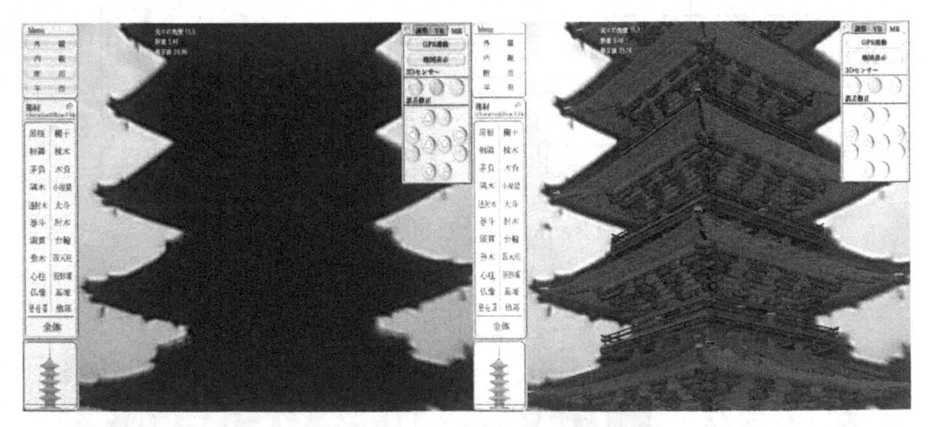

Figure 18. Gap with VR and an image of five-storied pagoda (left: before composition with VR and an image, right: composition with VR and an image).

6. Conclusion

Wooden framework architecture was reviewed in relation to the content and presentation method of digital archives of Japanese historical buildings. The definition of the components using a 3D model or framework construction was given. A description was given of the animated joints, et cetera, with Hi-Vision TV, and of the structure, which describes the framework construction interactively in a real-time simulation. Furthermore, the possibility was raised that the Tablet MR system could be used effectively as a simulation tool for historical buildings in on-site models. It is anticipated that in the interests of education about Japanese historical buildings, that the increased interactive potential according to users' needs will be increasingly necessary.

In future studies, many more simulations will be made and the system will be improved further until it is ready for use with interactive content, as an on-site model simulation tool. Improvement of the interactive content and on-site model simulation tool will contribute to the smooth advancement of e-learning related to historical buildings.

Future research directions may include device-related improvement, such as the accuracy of the 3D-sensor and to improve the portability of the GPS, the antenna must be miniaturised and made lighter. Moreover, in MR, a fault exists by which the object in front of the VR image disappears due to the overlapping of the VR image, so that it is displayed over the top of the object. Development of a technique to remedy this problem is greatly desired as it directly affects the user experience. In addition, it is necessary to prepare an environment in which a robust examination of the system functions can be performed.

Acknowledgements

The author expresses her appreciation to the following people. Regarding the contents of the work on To-ji, the start of the research was made by Professor Architect Toshiroh Ikegami who is the Director of a synthetic 'suggestion person,' which is still in progress. He is working with Kyoto City University of Arts, Toshiroh Ikegami Studio. Professor Architect Ikegami was General Director and Art Director of the Anatomy in Cyber Space-Toji Temple Exhibition 2003 Kyoto, 2004, China. Doctor Changsoo Byun of the Yeungnam College of Science and Technology performed excellent 3D modelling despite limited resources during the time of research as a doctoral student at the Kyoto City University of Arts. Doctor Byun performed all modelling of the structure of a five-storied pagoda, or animation creation alone from autumn 2000. After losing Doctor Byun, this project is continuing by another master course students and another university students. Doctor Wookhyun Yeo in Yonsei

University developed the VR system from August 2003 as a doctoral student of the Graduate School of Engineering, Osaka University at the time of this study. Masahiro Kawaguchi and Professor Tomohiro Fukuda developed the MR system. Further, I am thankful to the following people who cooperated with the research: students of the Environmental Design and Information Technology Laboratory, the Graduate School of Engineering Osaka University, doctoral course and master's course students of the Kyoto City University of Arts, the staff of Urbangauss, and graduate students of Kinki University's Junya Toda Laboratory. Finally, my deepest gratitude is offered to Professor Tsuyoshi Tee Sasada.

TEMPORAL CONTEXT AND CONCURRENT EVALUATION

Enhancing Decision Making at the Early Stages of Architectural Design with Mixed Reality Technology

JULES MOLONEY
The University of Melbourne, Australia

Abstract. A summary of the historical background to architectural drawing is presented, in order to locate mixed reality technology in relation to existing design traditions. From this background two ideas are introduced – temporal visualisation and concurrent evaluation – as the conceptual underpinning to the implementation of mixed reality technology at the early stages of architectural design. A second section reviews the taxonomy of mixed reality, and clarifies the requirements for a decision support visualisation environment. In conclusion, an approach being developed at the University of Melbourne in conjunction with HITLabNZ is outlined.

Keywords. Architectural Design, Time, Context, Mixed Reality.

1. Introduction

> I always recommend the ancient builders practice by which not only drawing and pictures but also wooden models are made, so that the projected work can be considered and reconsidered, with the counsel of experts, in its whole and all the parts.
>
> (Alberti, 1443–1452)

Drawing, pictures and physical models have been the mainstay of architecture for centuries, allowing what Alberti refers to as the counsel of experts. The last two decades have seen the wholesale take up of the computer as the primary means of visualisation and documentation for construction, yet arguably today's digital tools are conceptually little different from the drawing techniques invented in the Renaissance. The power of the computer has dramatically increased the speed of representation and the capacity to coordinate construction

X. Wang and M.A. Schnabel (eds.), Mixed Reality in Architecture, Design and Construction, 135–153.

information, but for most of the profession, little has changed conceptually. The artful architectural perspective has been developed into the photorealistic rendering, both considered in terms of a viewing angle that displays the design from its 'best side.' Computer aided drafting has enabled the coordination of cladding, structure and the plethora of pipes and wires, for the most complex of building requirements. The current shift to coordinating this information in three dimensions as a building information model (BIM), will again increase the speed and efficiency of the documentation process (Eastman, 1999). But as evidenced by such as Guarino Guarani, the architects of the baroque were capable of drawing and coordinating the construction of complex geometry (Wittkower, 1975).

Apart from speed of production, what impact has digital technology had on design methods and thinking – and how might the next generation of mixed reality technology extend current approaches? Currently two significantly new modes of practice have been facilitated by digital technology. The first is the shift from two-dimensional working drawings, to the direct linking of three dimensional computer files with computer numeric controlled (CNC) machines. As has been well documented in relation to pioneers such as Frank Gehry, the linking of CAD files to CNC tools enables the efficient production of non-standard geometry (Kolarevic, 2003). The promise of this approach is that it allows the mass customisation of building components, facilitating a break from the era of Fordist production that has typified architecture of the twentieth century.

The second significant impact is located at the other end of the production line – the crucial early stages where the primary design decisions are undertaken. There is a shift away from replicating analogue techniques, to realising the power of the computer as a processor of information. Rather than considering a 3D model in terms of refining a singular design solution, as one might with a physical design model, some designers are developing multiple solutions via parametric design techniques (Burry and Murray, 1997). The approach is based on conceiving a three dimensional computer model as a series of discrete but linked assemblies, so that changes in parts are propagated throughout the whole. If used during the production design stages, this facilitates the efficient development of the design details, but the more significant adaptation of the linked assemblies potentially occurs if this approach is used at the early stages of design. The geometry of a three dimensional model can be controlled at the 'meta level' by parameters, which propagate changes throughout the various assemblies that make up the whole. By manipulating these higher-level parameters, a wide range of potential solutions can be developed in relation to factors such as site conditions, surface to volume building efficiencies, or as the means to experiment with novel form. The parameters can be intuitively changed in real time by the designer to enable a range of solutions, in a digital form of 3D sketching (Figure 1).

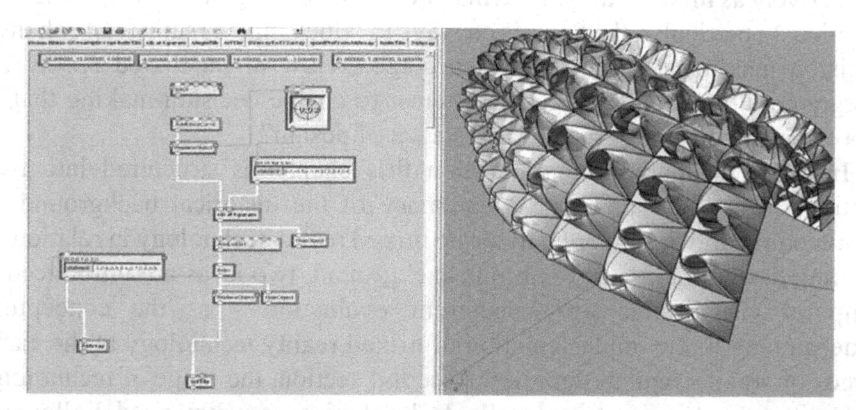

Figure 1. Interface linked to parametric form (Janssen and Krammer, 2007).

For many designers this facility to visualise in three dimensions in real time would be enough – the computer in effect fulfils the role as an efficient 3D sketch book, with the bonus that the relative accuracy of the computer model allows thorough evaluation of external and internal form and space. However, an extension of parametric design where the designer intuitively manipulates parameters is generative computing, in which the parameters are linked to systems that automate the generation of permutations and test these against the design criteria of the particular project. The criteria can be coded in relation to a functional brief, site factors such as sun shading or natural ventilation, or in terms of the predisposition of the designer for a particular design character or schema (Janssen, 2006). Typically, research in the use of generative computing has concentrated on functional performance, but in principle any design idea that can be described in terms of a range of parameters, can be codified to generate a wide range of solutions. These solutions may provide differences in degree, or, in an ideal scenario, unexpected outcomes evolve to offer differences in kind. The significance of this approach is that it shifts the working process from thinking in terms of manipulating a singular model, to the automated generation of multiple permutations. Design activity shifts from 'hands on' making, to the specification of parameters from which multiple forms are presented. Thus the designer, in effect, acts as editor of potential solutions that have been generated in relation to sets of parameters. The act of design occurs at the stage of parameter specification, but more importantly, during subsequent review stages where selection and editing occurs. Generative computing techniques in effect provide a wide range of tested alternatives, however, the design team have to select, edit and develop the preferred design.

These innovations – parametric and generative design – have the potential to transform practice beyond speeding up existing methods. The issue being explored here are the possibilities offered by the range of technologies known collectively as mixed reality, in terms of complementing these shifts in design practice. In particular, the focus is on ways in which mixed reality can enhance decision making at the early design stages. How might mixed reality be integrated into parametric design systems, to enable decision-making that is more considered at the crucial early stages of design?

In order to approach this question this chapter is structured into two sections. The first starts with a summary of the historical background to architectural drawing, in order to locate mixed reality technology in relation to existing design traditions. From this background, two ideas are introduced – temporal visualisation and concurrent evaluation – as the conceptual underpinning to the implementation of mixed reality technology at the early stages of architectural design. In the second section, the range of technology and approaches to using mixed reality technologies are summarised. Following the overview, is a description of an approach being developed at the University of Melbourne in conjunction with HITLabNZ. This project utilises a range of technology to explore the potential of temporal context and concurrent evaluation for the early stages of design. The research is predicated on the belief that new media and technologies ought not only facilitate existing modes of practice, but also engender new modes of designing. In our view, while innovation lies in expanding mixed reality technology for appropriation by design practices, the wider significance lies in extending intellectual traditions in representations and their power to encode, explain and transform our world.

2. Impaired Vision

The history of design visualisation for the bulk of the twentieth century can be directly related to the cannons of modernism. From its publication in 1932 until the mid 1970s 'The International Style' was the mantra for a period of 'impaired vision' in which architecture was "conceived as a thing in itself, as if it were the only building in the world, and designed from the inside out, in terms of an abstract, idealized conception of its functions, with no concessions to the landscape or cityscape around it" (Berman, 1988). Commiserate with this ethos, architectural drawings were abstract plans and sections delineating internal function and structure, supplemented with axonometric projections to show three-dimensional relationships as if they were describing engine parts. Elevations or perspective studies were identified with a morally suspect 'pictorial' view of the world. As has been well documented, the impact of this mechanistic view of architecture has had a catastrophic effect on cities world-wide and by the mid 1970s was widely discredited. The socio-cultural reasons for the failure of functional modernism are understandably complex and dependent on local circumstances, but one outcome was contextual relationships became of major concern. New theoretical texts such as 'The Concise

Townscape' (Cullen, 1961), 'Learning from Las Vegas' (Venturi *et al.*, 1972), and 'Collage City' (Rowe and Koetter, 1978) reasserted the importance of the appearance of architecture and how this is perceived in relationship to physical setting. These alternative theories stimulated and in turn were informed by visualisation that attempted to place proposals in context: photographic collage; the re-immergence of perspective drawing and sequential sketch techniques; the use of urban figure ground analysis; and the use of physical site models.

The hot-house for design theory and the interplay between ideas and drawing in the 1980s was the London Architectural Association school, particularly in the design studios of Bernard Tschumi and his student protégés. Inspired by Situationist theory, Tschumi's continuing aim is to reassert the importance of architecture as place for the unfolding of events (Tschumi, 1996). A key strategy was experimentation with alternate forms of architectural visualisation and his design classes at the architectural association were among the first to inform architectural drawing with a sense of occupation through time. The term 'narrative drawing' was coined to describe this radical approach to design visualisation in which drawings were developed as filmic story boards, mixing standard architectural conventions with photographs, and music or dance notation. The promise of narrative drawing and other temporal techniques such as Cullen's sequential sketches or earlier uses of video (Appleyard *et al.*, 1964) were overshadowed in the late 1980s by the widespread take-up of computer aided design (CAD).

Computer visualisation in architecture and urban design has evolved over the last twenty years as graphics performance has improved and the software has matured. CAD initially duplicated manual drawing conventions with the use of 2D plan, elevation and perspective views printed out and rendered by hand with pencil, ink or watercolour. The second generation of software allowed the design to be modelled in three dimensions, 'coloured' by the application of photographs of materials (a technique known as texturing), and lit by simulated light sources. Early attempts at these techniques were blocky and crude due to hardware limitations and lack of specialist knowledge. Typical architectural practices did not have the time or in house skills, and preparation of presentation drawings and animations were (and still are in many cases) outsourced to a CAD visualisation office. These digital specialists have replaced the traditional architectural illustrators of previous eras.

The increase in graphics power of the desktop computer, the development of more sophisticated and intuitive software, and the increased skill level of young graduate designers has lead to many contemporary practices being able to generate sophisticated computer graphics in-house. However, the ease with which complex geometry and surface qualities can be manipulated in three dimensions has meant 3D visualisation has become, to some extent, a victim of its own success. In the hands of many there would appear to be a predilection for abstract form-making undertaken in the vacuum of the 3D digital work-space, often exploring non-standard geometry with seductive but unrealistic surface and light qualities. Seldom are these experiments in form-making evaluated against the urban or rural context, until late in the design process. Even in a practice culture that emphasises contextual issues, design evaluation in a realistic mode seldom occurs until final client presentations. This deferral represents the time constraints within professional practice as much as it does particular design approaches, as while hardware and software have improved dramatically the scale and complexity of urban contexts still limits visualisation at the crucial early stages of design. There is no time to experiment with a range of viewpoints or camera tracks and lighting conditions are usually based on one idealised moment in time. The mode of presentation to clients becomes cinematic, computer visuals and animation utilised essentially as a marketing tool, with no ability for the reviewers to experiment with alternate camera positions or temporal contexts. The ubiquitous 'fly through' animation may give some appreciation of spatial sequence albeit from a non-grounded perspective, but there has been little advance in terms of evaluating design proposals in context *before* key decisions are made. Arguably the wholesale take-up of 3D digital modelling within contemporary architectural practice has seen a regression in terms of context evaluation, and a reinforcement of the paradigm by which architecture is conceptually and in many cases literally developed as an isolated object.

This brief review locates a gap in design practice in the area of design visualisation. Despite the recognition that architecture affects and is affected by context, design evaluation at the early stages occurs either through 2D drawing conventions developed in an era of functionalist modernism, abstract physical models, or in the contextual vacuum of 3D CAD. Moreover there are no examples in which contexts are considered over time. The advent of Virtual Reality (VR) and screen based Virtual Environments (VE) has allowed experimentation with design in real time within a virtual site context. However, despite the time and care that goes into developing virtual environments, they remain a virtual simulation-based on single snap-shots in time. By contrast, the dynamic quality of context of all design projects can be readily experienced. Each hour, day and month is unique, being affected by changing weather, usage patterns and local events.

3. Temporal Context

The foregrounding of time in relation to context in design is not a new idea, but one that has struggled against ingrained traditions that privilege the architectural object over context and the lack of efficient representation techniques. It is self-evident that architecture must perform within a dynamic context, but as outlined in the review of design visualisation above, in practice, context is considered at best in relation to a snapshot in time. The idea of evaluating context in relation to time is developed here into two interrelated aspects: (1) that architecture is experienced by the body in motion as a *spatial sequence* over time; (2) that architecture must perform functionally and in socio-cultural terms in relation to a *dynamic environment*. The simile of film provides an illustration of these aspects and how they interrelate.

Filmic experience is based on both the mobile camera that tracks a scene, and the stationary camera that presents a fixed point of view and captures events as they unfold within the frame. In film, spatial sequence is usually compressed as a temporal montage of moving points of view and it should come as no surprise that the pioneer of this technique, Sergei Eisenstein, was trained as an architect (Eisenstein, 1949). The second filmic approach relies on a static camera that captures action through the real time unfolding of events, via temporal cuts that return to the same scene, and in some instances, time-lapse techniques that literally capture the dynamism of the environment. The experience of film, like architecture, is a montage of spatial sequences that are conditioned by a dynamic environment. While we cannot experience one without the other, for even while physically stationary the eye is continually shifting direction and focal length, it is useful here to consider them independently as each provides distinct technical challenges for mixed reality systems.

The first aspect of the temporal context – spatial sequence – has a long tradition in terms of how architecture is experienced. Bois and Shepley's 'A Picturesque Stroll around Clara-Clara' (1984) traces a genealogy of the 'peripatetic view,' from the Greek revival theories of Leroy, the multiple perspective of Piranesi, Boulée's understanding of the effect of movement, to the Villa Savoye where architecture is best appreciated, according to Le Corbusier 'on the move' (Bois and Shepley, 1984). But how has this tradition been supported in terms of design media and techniques? Le Corbusier and generations of twentieth century architects have relied on plan and section, supplemented by cryptic perspective sketches to organise movement. The close observation of existing building and cities coupled with the long apprenticeship of formal training and as an office junior, honed the architect's eye to imagine the experience of occupying and moving through drawings. The possible paths through plans often traced out as faint circulation lines – an internalised playing out of spatial sequence within the imagination of the designer. Robin Evans eloquently describes this imaginal art, the translation from projective drawing to constructing the experience of architecture through time.

> Design is action at a distance. Projection fills the gaps; but to arrange the emanations first from drawing to building, then from buildings to the experience of the perceiving and moving subject, in such a way as to create in these unstable voids what cannot be displayed in designs – that was where the art la.
>
> (Evans, 1995)

Evan's research of the role of drawings in relation to the development of architecture is seminal. The relationship between developments in technique, built form and experience, the ability to describe and imagine form on paper, is presented as an alternate history. In a summative diagram, Evans describes the activity of drawing as a tetrahedron where in one node the idea forms as an internalised 'picture.' It is then explored via the architectural sketch, 'sold' to the client via rendered perspective and realised by orthographic drawings. This diagram makes explicit the separation between the creative activity of architects and the ultimate result – architecture experienced by the body in motion. Evans' thesis is an admirable *history* of architecture and his diagram a useful summary of architectural activity in which drawing is the dominant activity. It is however, an exclusive account: physical models are accommodated within Evans' thesis as intermediate modes which require translation via projective drawing. The observation that since the advent of cinema there have been successive generations whose experience of vision has been dominated by the mobile image is not addressed. Additionally, although Evans' book 'The Projective Cast' was published in 1995, he also neglects to mention that by the early 1990s, computers were increasingly prevalent in architectural studios. Therefore, as an investigation of the influence of geometry via media, it is a history that stops with the advent of photography. What changes, if anything, by projecting the digital into Evans' diagram? Would it be possible to revise it, to close down the distance between representation and building? A review of research in architecture and computing in the mid nineties reveals that there were three fledgling procedures that were transforming the embedded practice of projective drawing: parametric design, immersive editing and computer aided construction (Moloney, 2000).

This chapter started with the proposition that parametric design and computer-aided construction were the two most significant developments in architectural design, and arguably these are now reasonably well understood. What was meant by the phrase 'immersive editing'? At the time I had just shifted from a professional to an academic career and was, like many, inspired by the development of virtual reality technology. The hype was of immersive virtual worlds where the designer could potentially occupy the digital model and design from the point of view of occupation. Unfortunately, it seemed that only those who could afford the latest SGI machines could test this idea. However, thanks to the suggestion of an insightful student, we embarked on a series of design studios that used video game technology to explore the idea of immersive editing within low cost screen versions of virtual reality. We deliberately used video game authoring technology that allowed simultaneous occupation by multiple users. This technology also enabled the visualisation of large terrains and cityscapes with dynamic lighting in real time. The focus was on designing both from within the architecture and exploring the impact of the design on the virtual site. Students could design in real time in plan, section or perspective in a highly rendered world, while the multiple user functionality allowed designers and critics to share and critique work as it was being conceived. To encourage this dialogue, PHP chat room functionality was incorporated to allow the asynchronous posting of comments. The implementation of the collaborative virtual environment is illustrated below (Figure 2).

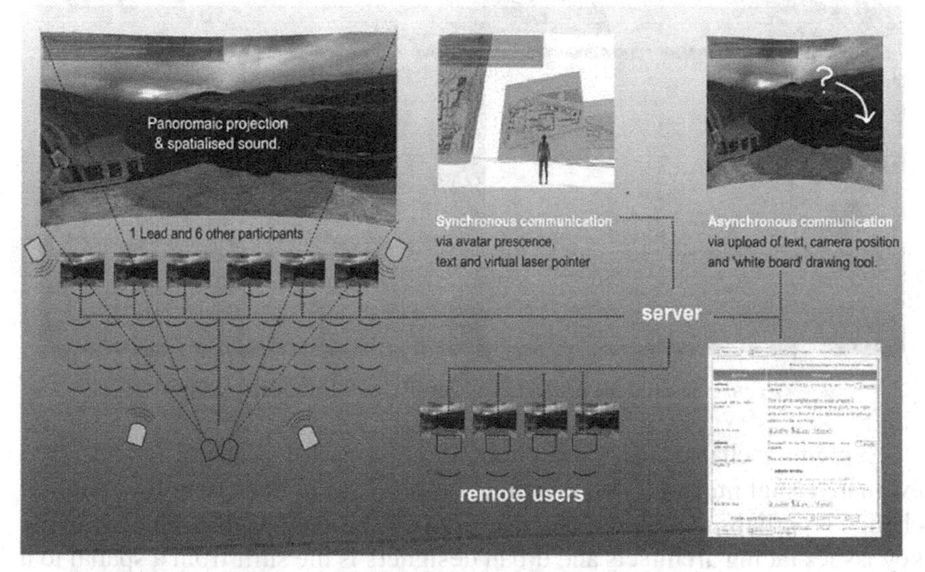

Figure 2. Collaborative virtual environment for design (Moloney, 2002).

Such visualisation, based on spatial sequence and the ability to freely change camera position and orientation, represents one aspect of the temporal context. As revealed in these studio trials, the free camera movement allows the designer to discover any unanticipated impact of the design in relation to the virtual context and a rich understanding of the formal qualities of the design as experienced in terms of sequence over time (Moloney and Amor, 2003).

The second aspect of the temporal context is that perception of the built environment and the performance requirements of buildings vary dramatically over time. In terms of perception, the obvious, but seldom exploited fact is the continual transformation of architecture in response to changes in light and moisture. A recent design studio undertaken at the University of Melbourne focused on the potential of architecture to exploit this perceptual change using time-lapse video taken over daily and weekly cycles. As the example below illustrates, geometry, reflectivity and opacity can be manipulated to transform appearance over time which can then be evaluated against high fidelity video context, albeit from a fixed viewpoint (Figure 3).

Figure 3. Student project, University of Melbourne, 2006.

Exploiting changes in atmospheric conditions for aesthetic effect or to evaluate visual impact is one outcome, but the idea of designing for temporal change is perhaps more important in terms of functional evaluation. One of the key issues facing architects and urban designers is the shift from a spatial to a temporal understanding of urbanity, and the need for 'designers to create appropriate flexible environments permeable to constant and rapid changes' (Echeverri, 2005). This view concurs with Tschumi's proposition that architecture is a geometric container awaiting, and transformed by, the event of occupation.

In an attempt to evoke the constant shifts in activity during daily and seasonal cycles, a second series of studios based on immersive editing was undertaken, which explored the potential of sound to convey the atmosphere of occupation over multiple time scales. The best of these brought 3D worlds to life (Harvey and Moloney, 2005). As one traversed through the architecture, snippets of conversation, passing traffic and footsteps evoked changing activities, reinforced a sense of materiality and the life of the city. These aural memory triggers were synchronised with shifts in ambient lighting and passing shadows, to reinforce the perception of architecture being experienced both in terms of a spatial sequence over time and in relation to a dynamic environment.

4. Concurrent Evaluation

The above discussion emphasised qualitative experiences of architecture over time. Ideally, evaluation of a design proposal can occur where qualitative aspects can be evaluated alongside quantitative data, such as constructional efficiencies and environmental performance. The simultaneous evaluation of the qualitative and the quantitative is captured by the phrase concurrent evaluation. Typically, these are separate conversations, with designs often evaluated in terms of environmental performance or efficiency of construction after key conceptual ideas are 'locked in.' Current thinking acknowledges that architectural design must perform in an expanded field in which performance in socio-cultural and aesthetic terms should be considered alongside functional performance. However there is a lack of methodologies and fundamental research to support concurrent evaluation of aesthetics and performance. Computer aided architectural design (CAAD) evolved during a period in which architecture was championed as a form of design science using the analytical approach known as design methods (Glanville, 1999). The legacy of this is that design is considered as a form of problem solving, hence the development of many computational approaches to generating solutions in relation to functional performance, and decision support based on technical analysis.

By comparison, there has been little research in providing decision support for the qualitative assessment of the formal properties of design, particularly in relation to context e.g., elegance of massing, profiles, proportions, the interplay between solid and void, articulation of materials and other aesthetic properties of design composition. From a scientific perspective, this is perhaps understandable as functional performance can be measured, charted and graphed but qualitative assessment is not readily amenable to calculation or 'proof.' Primarily it involves subjective decision-making with designers, clients and stakeholders discussing the formal merits of one design over another. Again, from a scientific stance, this subjective, discursive method is problematic – for example, assessment of a design in terms of environmental performance is measurable but how can we 'compute' aesthetics or visual impact? Proportional systems or shape recognition techniques can be used to determine objective qualities, but this is usually based on a gestalt model in which perception is explained in terms of neutral cognition of figure ground relationships (Gero, 1999). While there is still much more research to be undertaken within psychology, the current agenda is to consider perception a complex interplay between the full range of sensory inputs, memory and in the case of architecture the local socio-cultural and environmental framing of the architectural design (Bonnes *et al.*, 1995). If objective measurement is improbable, mixed reality systems offer the potential for the simulation of designs in a multisensorial and temporal context that enhances and aids subjective evaluation of designs prior to construction. In this context, mixed reality technology offers the potential for the concurrent evaluation of quantitative and qualitative attributes of design options, in relation to an interactive real-time parametric model, or what has become known as a 'digital prototype.' Figure 4 illustrates how the two ideas – temporal context and concurrent evaluation – can potentially extend the current use of parametric digital models to provide a holistic prototyping environment.

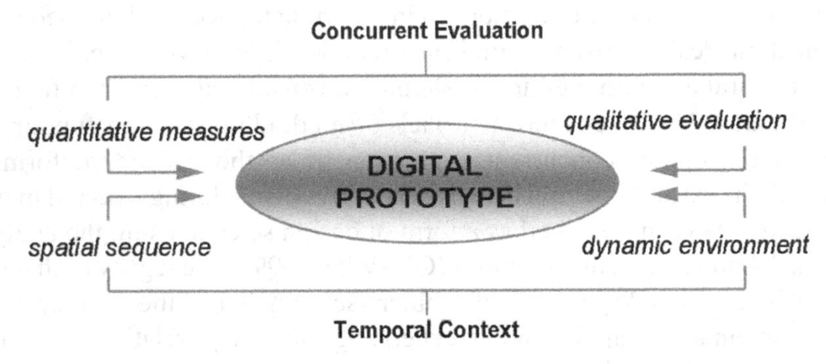

Figure 4. Enhanced digital prototype (Moloney, 2008).

The diagram extends the typical understanding of a digital prototype as a discrete building model evaluated in the contextual vacuum of the engineering design interface, to one where the qualitative can be evaluated alongside the quantitative. This is both in terms of a temporal context, where evaluation is based on moving through typical spatial sequences, and the simulation of a dynamic environment. As discussed above, these concepts have been developed over the last ten years and partially tested through a series of design studios. The results have been promising, but there are limits to adopting videogame technology. We have found mixed results over the years, with the quality of the contextual framing being dependent on software skills that are above that found in typical design studios – convincing texture mapping, lighting and soundscapes take considerable time to produce the photorealistic outcomes seen in contemporary videogames. In addition, no matter the skill level, the virtual environment will always be an interpretation of the actual context.

The promise of mixed reality, in terms of the focus of this chapter on the early conceptual stages of design, is that the combination of the real with the digital will address these production problems and extend the agenda from evoking context to one closer to actual experience. The next section surveys the available technology and evaluates applicability for use at the early stages of design and in the process articulates a distinction between the activity of designing and the collaborative process of the design review.

5. Mixed Reality for the Early Stages of Design

The definition of mixed reality most frequently encountered is based on the taxonomy of Milgram and Kishino (1994), in which they classify visual display environments that allow the combination of 'virtual' and real space. Rather than articulate hard boundaries between fully synthetic digital space and the video display of real environments, they develop what they term a 'virtuality continuum.' The location of display environments along this continuum is determined by "the extent of knowledge present within the computer about the world being presented" (Milgram and Kishino, 1994). At one end of the continuum is virtual reality, where all the information required to produce the contents of the display environment is known (geometry, location, surface, etc.), while at the other is video projection of a physical environment where all that is known (by the computer) is that it is a video file. This allows a distinction between the types of displays that combine the real and the synthetic, of which there are two primary types – augmented reality (AR) and augmented virtuality (AV). Augmented reality is video of a real world environment onto which is superimposed digital models, experienced within a CAVE (Computer Automatic Virtual Environment), HMD (Head Mounted Display) or screen. In this case there is minimal information about the environment being displayed – all that is required to align the digital model, is the position and orientation of the video camera. Augmented virtuality on the other hand, is predominately a synthetic world into which is incorporated video of real world objects or scenes. Within the primary distinctions of AR and AV are a number of other approaches that have been classified along the continuum from real to synthetic worlds, including amplified, mediated and virtualised reality (Schnabel *et al.*, 2007).

The objective in this review is to locate opportunities along the virtuality continuum for a particular application – the concurrent evaluation of digital prototypes in a temporal context at the early stages of architectural design. What mix of technologies will enable design ideas to be evaluated in relation to a dynamic context, from multiple motion paths, and at the same time allow the superimposition of functional performance data? Perhaps more importantly for design practice, how might these technologies be implemented in a studio design environment? At present, the majority of the prototypes and case studies have been developed as university research projects that demonstrate technical feasibility, with scant regard for how the technology may be used and integrated with existing design practice.

The survey of Schnabel *et al.* (2007) usefully clarifies the issue of suitability for specific activities. Their review evaluated technology along the virtuality continuum in terms of two factors (1) correlation between perception and action (2) level of interaction with real artefacts. The first factor is based on human computer interaction research that demonstrates the more we can use our every day motor activities in interacting with virtual objects, the better the performance (Aicher, 1997). For example changing the location of a digital model by physically moving a marker, enables more intuitive interaction than moving the model by mouse or keyboard commands. The second factor is based on research in activity theory that takes the perspective, "that the computer is just another tool that mediates the interaction of human beings with their environment" (Kaptelinin, 1996). In contrast with earlier cognitive theories of human computer interaction, activity theory takes into consideration the impact of traditions with non-computer tools and the social context in which digital tools will be used. 'Which tools, other than computerized tools, are available to the user? What is the structure of social interactions surrounding computer use?' (Kaptelinin, 1996). These factors are of particular importance to architectural design at the early stages. Hence, before evaluating the possible mix of technology to achieve concurrent evaluation of designs against a temporal context, it is worthwhile reflecting on tools in existing design practice and the need for social interaction during design.

> discussing, arguing, negotiating, forming consensus, trying out ideas and getting reactions, identifying and resolving conflicts, and reaching shared understandings and agreements. It's a person-to-person, social process – not just one of solving technical problems and producing documentation. That's what practical design is mostly about.
>
> (Mitchell, 1995)

William Mitchell's keynote address at CAAD Futures'95, highlighted that design review and decision making is reliant on conversation and negotiation, usually involving a range of representations – performance data, drawings, physical and computer models. Despite advances in distance-based collaboration technology, crucial design decisions are still undertaken in face-to-face review and discussion, 'trying out ideas and getting reactions.' This observation enables a clarification of the potential role of mixed reality technology at the early stages of design.

We should be clear on the distinction between the individual act of *designing* – the formation of early ideas on paper, and with physical and digital models – and the process of *design review* in which these sketch designs are evaluated. In the actual idea formation stage, there is a strong argument that working with abstractions rather than a fully rendered context is more productive (Do and Gross, 2001; Herbert, 1993). There is of course no normative design method that captures the range of approaches individuals use to generate ideas. But as a general theme, whether developing ideas with paint, charcoal, graphite, Photoshop, Sketchup, Maya or other 3D modelling software, there is a tendency to deal with abstractions rather than with detail. Typically this is a solitary, reflective activity, the individual designer developing ideas as an internalised conversation with media – the "mark, interpret, mark" cycle identified by Herbert in relation to drawing practice (Herbert, 1993). Moreover individuals have favourite tools, hence having to learn new computer interfaces may be counterproductive at this idea conception stage. The design review by contrast is a group activity, in which design options, often from a number of designers, are compared and discussed in relation to a range of issues. Arguably, it is at this point of the early design stage, that the ideas of concurrent evaluation within a temporal context are most productively introduced. The individual ideas can be evaluated side by side, against the context in which they have to perform, allowing discussion of qualitative and quantitative performance within a range of time scales.

What are the requirements of a mixed reality display for group design review in a design studio? In summary the technology should, at least, be capable of the following.

- Shared Display Environment: Provide a display environment that allows social interaction, natural dialogue and access to documents, drawings and physical models.
- Navigation: Allow real-time navigation in three dimensions and/or multiple camera paths through the environment.
- Dynamic Environment: Ability to change the temporal scale against which decisions are made (daily, seasonal cycles or longer time scales).
- Concurrent Evaluation: The facility to display performance data alongside the design visualisation.
- Interactivity: For design reviews, relatively limited model interactivity is required – the facility to easily swap design options, move, rotate and scale translations would suffice.
- Lighting: Allow the import of models from typical design software and automate the process of lighting the design model in relation to the context.

What mix of technology might meet these requirements? For a start, we can negate one end of the virtuality continuum – virtual reality systems are not appropriate as full immersion in a synthetic world removes the participant from the physical space, working against the requirement for natural interaction and access to supporting drawings or physical models. The choice would appear to be between augmented virtuality (AV) or augmented reality (AR) and their variants along the virtuality continuum, each of which may be appropriate dependent on the specific design task. AR allows the superimposition of a digital model into an actual context and would be appropriate for typical architectural designs, where the scale of the designed object is significantly less then the context. Rather than model and texture the whole context (a time consuming and specialist skill), this could be represented by a site video, with the design model aligned with the camera view. Conversely, with city planning and urban design, the scale of design content may be more than the context, and would suggest AV would be the appropriate approach. Take for example the evaluation of a 20 year urban design expansion in relation to an historic quarter. In this case embedding video footage of the historic quarter within the urban design model may be appropriate. Returning to Milgram and Kishino's taxonomy (1994), the choice of technology may be determined by the question – what is the extent of knowledge needed within the computer about the world being presented? In the case of large scale planning and urban design, which may be staged over many years, the computer would need information on most aspects of the display environment, and hence AV would be appropriate. In the case of design at the scale of a singular building or a city block, in order to display the design options in context, the amount of information would be comparatively small, which according to the taxonomy suggests AR would suffice.

Let us consider the latter more general case for architecture – from the scale of a single building to that of a city block – and some of the variants of AR approaches that have been used in architecture to date. Azuma provides a robust definition of augmented reality as having the following attributes: they combine the virtual and the real; the virtual is three dimensionally aligned or 'registered' with the real; and the systems allow interactivity in real time (Azuma, 1997). The visualisation environment may involve the use of head mounted displays or portable screens where the user is physically in the space – known as mobile AR, or a second mode where the virtual is superimposed on pre-recorded video and visualised with stereo projection systems – non-mobile AR. There have been several experiments in the use of head mounted displays for design visualisation, but these have been technology trials outside the design domain, or pilot studies which have had minimal impact on theory or practice: excellent technology trials have been undertaken at the University of South Australia with an outdoor AR CAD system (Piekarski and Thomas, 2003c), while Feiner *et al.* (1997) explored the use of mobile AR to visualise historic buildings on location; design studio tests have been undertaken at architectural schools (Kuo *et al.*, 2004). HMDs have also been used with physical city models (Seichter, 2003a) but these substitute an abstract computer model with an abstract physical model with arguably, little improvement in context visualisation. Moreover there is a growing consensus that HMDs are not conducive to design as a social process, and the importance of integrating traditional media with computer visualisation (Dave, 2003). Design reviews require dialogue between several participants, and a screen system rather than an HMD, allows more natural conversation and also the simultaneous evaluation of reports, drawings and physical models. It would appear none of the above approaches meet the specific requirements of visualisation environment that allows concurrent evaluation against a temporal context, in a manner that addresses the social interaction that occurs during a typical studio review.

6. One Approach to Decision Support: StrollAR and Video-datAR

Below is described an approach to provide such decision support at the early stages of architectural design, that is currently being developed at the University of Melbourne in collaboration with HITLabNZ. The project uses a range of mixed reality technologies to meet the list of requirements described in the previous section.

- Shared display environment: Adopt a screen-based approach to allow studio design review using stereographic projection, complemented by a mobile screen that allows on-site evaluation and consultation.
- Navigation: Provide both real-time navigation and multiple camera paths that align the digital model with pre-recorded site video. This requires the combination of AR and AV in the one visualisation environment. The AV mode would consist of a low polygon model of the environment together with the 'sky box' approach used in video-game production, where the background is rendered as a 360 degree panorama. When switching to AR mode the camera would align the design model in relation to either streaming video (mobile AR on site) or in relation to pre-recorded camera paths (non-mobile AR in the design studio).
- Dynamic Environment: Develop a database of site time-lapse video taken from key viewing points. Link design models to environmental performance software that considers a range of time scales. In real-time (AV mode) animate lighting and skybox relative to range of time scales.
- Concurrent evaluation: Develop an export module that converts the design model into a format compatible with environmental performance software. Import the performance data and store in local database to enable continuous update of performance as different options are evaluated/different time scales are considered.
- Interactivity: Enable users to swap between the navigation modes (real-time or via pre-recorded camera paths) on-the-fly, swap camera paths and alter video playback speed and direction. Design models should be able to be swapped in and out on-the-fly, and entities should be editable in terms of translation, rotation and scaling.
- Lighting: Develop a lighting export module that automates the lighting of design models in external graphics animation software in relation to the pre-recorded video.

An initial prototype based on the above requirements has been developed (Moloney, 2006). This consists of StrollAR (Figure 5, left), a mobile screen based system that is initially used to survey the site, collecting a database of motion paths and time lapse studies. This can also be used to provide on site evaluation, where the design model is updated in real time in relation to camera position and angle. StrollAR is complemented by a second system that is intended for use by designers in a studio situation. As illustrated below Video-datAR is a multi-screen projection display linked to three databases: motion and time-lapse video of the site, a 3D-design model database, and a building performance database that displays information in a graphical format.

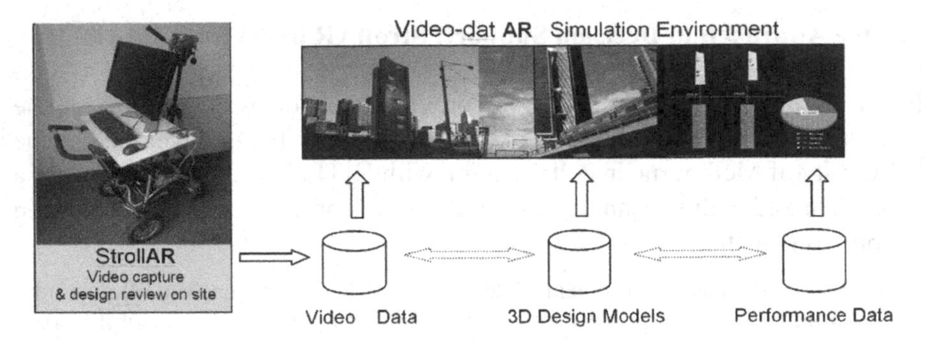

Figure 5. Mobile StrollAR and non-mobile Video-datAR (Moloney, 2007).

Designers can develop ideas in their favourite application or design directly in the virtual environment linked to the video database. Designs can then be tested in context with the video camera paths and time-lapse studies. Where the pre-recorded camera paths are insufficient, the user can switch to the AV mode and explore the design with a free camera. The 8:3 aspect ratio of the site video counters the 'tunnel vision' issue of perspective viewing and allows the wide screen format of a two or three screen projection facility to be used effectively. A preliminary interface has been developed to allow the concurrent graphic display of key environmental performance data generated by an off-the-shelf analytical application. The scale and wide field of view of the projection maximises engagement, yet allows a valuable 'distancing,' and the simultaneous evaluation of models and drawing. The capacity to shift focus from a semi-immersive screen to reports, sketch books, orthogonal drawing and physical models encourages the 'reflection in action' acknowledged as most conducive to creative design (Schön, 1983). Once design options have been agreed upon and developed in the studio, the mobile StrollAR can then be used to communicate and discuss the outcomes with a wider range of stakeholders on the actual site.

7. In Conclusion

This chapter has deliberately emphasised the historical traditions of representation in architectural design, with the aim of stimulating debate on the relationship between mixed reality technology and current practice. Architecture willingly engages with new digital technology, but new tools are usually absorbed into ingrained traditions of conceiving and evaluating design. One observation is that current use of 3D design software continues a modernist conception of architecture as 'machined object' devoid of context. Parametric design, BIM and computer aided construction are valuable innovations that transform the use of the computer beyond that of an efficient 2D drafting tool.

However, the context against which the digitally conceived design is evaluated (if at all), usually undertaken with abstract block models that allow basic sun shading and sightline studies. There are of course exceptions to this general trend – one hopes for example, that the tradition of sketching ideas on site may still be alive as a traditional form of 'mixed reality.' However, the image of the designer spending significant time on site, closely observing context with sketchbook to hand is a romantic one, out of step with the demands of contemporary professional practice. Moreover, it is an image that privileges the qualitative aspects of design and the unfortunate habit of evaluating functional performance after key design decisions have been 'locked in.'

As a means to clarify some advantages for the use of mixed reality technology in relation to the early stages of design, two ideas have been put forward for discussion: temporal context and concurrent evaluation. These are summarised by way of a diagram (Figure 4) that extends the use of a digital prototype. These concepts have been developed after a lengthy period of experimentation with virtual environment software via university design studios and academic research. They will require detailed evaluation in relation to current professional practice, if the motivation to shift 'object based' thinking is to be successful. It is to this end that I have argued that an important distinction needs to be made between the act of designing, and the design review.

Architects already have a range of tools and approaches that have been tried and tested over generations, and these are now complemented by parametric and generative design techniques enabled by advances in design software. It is debatable how much mixed reality adds to the act of designing at the early conceptual stages, where the most intuitive design interface is still eye-hand-graphite. It is argued here that one potential for mixed reality is to support decision making during the design review stage, where designers, clients and stakeholders evaluate a range of options. This is where the crucial decisions are made that set the trajectory for detailed design. Aimed at this particular design stage, a list of requirements have been outlined. These are intended to meet the needs of a design review that allows the concurrent evaluation of qualitative issues (such as massing, visual impact and materiality) alongside quantitative data on functional performance – all of which evaluated are within a range of time scales. Comparing these practice requirements against the technology presented within the mixed reality continuum has enabled the outline specification of technology to provide decision support at the early stages of design. A prototype system has been undertaken, that will be further developed and tested in practice-based case studies as part of ongoing research undertaken by the University of Melbourne and HITLabNZ.

4 MIXED REALITY IN CONSTRUCTION

KEY AREAS AND ISSUES FOR AUGMENTED REALITY APPLICATIONS ON CONSTRUCTION SITES

PHILLIP S DUNSTON AND DO HYOUNG SHIN
Purdue University, USA

Abstract. In this chapter, the potential application areas for AR to support construction phase activities are explored. As a medium for accessing information to support site operations and tasks, AR can benefit three primary categories of activities that occur routinely in the construction phase: (1) building and inspecting, (2) coordination, and (3) interpretation and communication. This chapter also discusses AR system technical issues regarding displays, tracking, and calibration for characteristics of construction tasks and the construction site. These technologies should be developed with critical consideration for the characteristics of construction sites if successful adoption of AR is to occur.

Keywords. Application Areas, AR Medium, AR System, Construction.

1. Introduction

The primary benefit of Augmented Reality (AR) is that it enables delivery of computer-mediated contextual information to the user that may not otherwise be readily accessible. With the trend toward more extensive use of computers in the development, capturing, and transmission of data, information, and knowledge, AR presents itself as a uniquely innovative option for supervisors and workers to interface with computers in more intuitive ways. AR technology attempts to deliver information as seamlessly as possible to facilitate improvements in decision making and thus performance. As such, AR's potential for impacting field performance during the construction phase of AEC projects is worthy of investigation. In this chapter we explore the potential application areas for AR to support construction phase activities. We look at how AR may provide field personnel with visual aids to perform work tasks more efficiently.

X. Wang and M.A. Schnabel (eds.), Mixed Reality in Architecture, Design and Construction, 157–170.

2. AR Technology Presence on the Construction Site

Before addressing specific applications of AR on the construction site, it is worthwhile to visualise what AR would look like on the modern construction site. First of all, AR-based devices for the construction site would be of light weight and small size to facilitate mobility. Wearable computers or ubiquitous environments in which servers deal with computing processes so that even wearable computers are unnecessary would be appropriate platforms for computer processing. Displays for AR systems on the construction site should be also lightweight and small. Head-mounted displays (HMDs), hand-held displays, or tripod-mounted displays would be the array of alternatives one should expect to find.

AR systems for construction sites would also cover wide ranges of space while achieving and maintaining sufficient levels of accuracy for various requirements. Because individual tracking technologies, such as magnetic, optical, ultrasonic, inertial trackers or GPS, have accuracy limitations and specific vulnerabilities in the construction site environment, tracking would perhaps be accomplished by more robust hybrid tracking technologies (Dunston and Wang, 2005). For example, a hybrid technology similar to the InterSense VisTracker which combines optical and inertial tracking would be appropriate for a truly mobile system. The system's reliance upon numerous tracking markers, however, constitutes a limitation upon which improvements would have to be made.

AR-based devices for the construction site would be easy to operate. Workers focusing on their tasks would not have much tolerance for devices that are difficult to operate. Device interfaces would be as simple and mobile as possible. Touch screen inputs, small simplified keypad configurations, and wrist attachments would be common features. On the other hand, the haptic interfaces commonly employed in VR environments or tabletop Mixed Reality applications would be rare for AR applied in the construction phase because the users of AR during construction would be more focused upon handling real structures or elements of the structures.

This brief detour in our discussion provides an image of the readily visible changes on the construction site after AR technology has been implemented. It does not immediately strike one as being dramatically different than what is seen today. Construction contractors already bring their PDAs and notebook computers to the construction site, and laser measuring devices and GPS-based systems for locating structures and guiding equipment are becoming less and less of a rarity. So perhaps this vision is not so far away as one might first think. The key to this seamless, almost quiet adoption will be addressing and resolving key hurdles in technology development and in selecting tasks that are most suitable for exploiting the benefits of AR.

3. Potential Application Areas

A growing number of technical feasibility studies of AR in construction have shown the potential of AR in the construction phase. For example, Webster *et al.* (1996) created an AR prototype for architectural assembly that provides users with AR guidance for assembling a space-frame structure. Roberts *et al.* (2002) presented an AR system that superimposes a virtual representation of subsurface utility systems, such as buried pipes and cables, onto the real outdoor environment. Although their system prototype was demonstrated as a tool for revealing buried utility locations as inventoried and documented in a municipal database, the application to field construction is evident. Their system might enable users to interpret easily mapped data of subsurface systems and thus facilitate the avoidance of accidental damage to those underground systems during excavation works. Behzadan and Kamat (2006) built an AR prototype designed to place virtual objects at desired outdoor locations for construction operations simulation. This application aids in the visualisation of the specified stage of a proposed structure or site operation, thus reducing susceptibility to mistakes in interpreting plans or designs.

However, as Shin and Dunston (2008) pointed out, realisation of these uses of AR technology requires not only demonstration of technical feasibility but also validation of its suitability. The potential application areas for the use of AR identified from their study can be classified as follows:

- Building and inspecting
- Coordination
- Interpretation and communication

Based on Shin and Dunston's study (2008), we will survey these application areas of Augmented Reality in the construction phase. While that work specifically considered the construction tasks routinely found in industrial construction, we believe that the same fundamental opportunities apply to other types of projects.

3.1. BUILDING AND INSPECTING

From the early exploration of AR for construction tasks, it has been envisioned that a major potential application of Augmented Reality technology is that of a visual aid to guide the actual building of the project and then inspection of the product. Feiner *et al.* (1995) and Webster *et al.* (1996) were among the first to actually demonstrate its practical use to the construction research community for construction assembly and maintenance inspection. In this kind of application, AR can be seen as an improvement over other visual aids used to provide reference points or lines, leveraging the growing availability of digitally generated design information. Noteworthy examples of potential applications of Augmented Reality for these purposes are layout, excavation, installation, and inspection. Following are brief explanations of each.

3.1.1. Layout

In the layout process, it is necessary to perform either a distance or angle measurement, or both, from a known starting point in order to decide the positions of new reference points on the site. To determine correct reference points on the site, time is required to set up a measuring device and read precise measurements. However, virtual reference points marked in the digital design model and then superimposed accurately onto the worker's view of the site may free workers from the task of measuring to establish the reference points on the site (Figure 1). Workers can identify the positions of reference points easily and mark them on the site by simply observing the rendered virtual reference points, and with an appropriately designed display (viewing device) may not even need to physically mark reference points or lines because they would be able to see virtual references as they need such guidance.

Figure 1. A conceptual view of AR overlay for layout.

3.1.2. Excavation

In an excavation process, equipment operators traditionally achieve the design grade level by referring to information coded on grade stakes typically spaced at least 15 m apart and must interpolate the excavation grade levels between the staked locations. However, these stakes are often run over by equipment during the course of excavation, creating gaps in the grade information to which the equipment operator must have access. It is both time-consuming and expensive to replace stakes. After excavation, surveyors are also required to check the finished grade to assure compliance with design requirements. This iterative procedure is labour-consuming, especially for the surveyor(s), and requires the surveyors and the equipment operators to cooperate closely with each other (Baertlein *et al.*, 2000). Alternatively, a presentation of the virtual excavation target area, perhaps in stereo view, and the desired excavation grade level that are rendered onto the work site in actual scale, location and orientation may allow equipment operators to more intuitively and to achieve the design intent without interruption to the excavation task. They can excavate the ground by simply following the spatial cues provided by the projection of virtual 3D design information onto the work site scene as illustrated in Figure 2. Except for initial their involvement in defining the site coordinate system and its base references, this application of AR may eliminate the necessity of on-site surveyors.

To enable equipment operators to confidently and easily identify the desired excavation grade level with this AR view, the AR overlay of 3D design for excavation should provide a full depth cue. This full depth cue may be achieved by the stereo view, or if that is deficient, the interactive occlusion of the virtual 3D design and the real ground surface. For example, if the desired excavation grade level is below the actual ground surface, the virtual 3D design will be occluded by the actual ground surface. If the desired excavation grade level is above the actual ground surface, the virtual 3D design will occlude the actual ground surface.

The authors acknowledge that the excavation application would be in competition with the trend toward automated machine guidance (AMG) which involves the use of 3D terrain models of site designs and 3D position tracking technologies, such as GPS or laser (Jonasson *et al.*, 2002; Hannon, 2007).

Figure 2. A conceptual view of an AR overlay for excavation.

Careful thought must be employed to assess precisely how AR might be utilised to complement rather than compete with such technologies that have already become commercially available. Specifically, situations would need to be identified where actually seeing the AR scene adds to the quality of the finished product. One such example might be in remote control of construction equipment that must be operated in hazardous environments such as the well-known Chernobyl nuclear facility or near seismically active areas such as the Mt. Fugen volcano in Japan where teleoperated equipment and remote vision systems were actually used to construct canals (Oloufa *et al.*, 2003).

3.1.3. Installation

The installation process may reap benefits from AR as well. When accurate and precise installation of a structural element or equipment item, is required, a simple target location will not suffice. For example, a structural steel column requires not only placement in a specific location, but also a critical 3D orientation in terms of its vertical alignment, the latter not being achievable simply by knowing the target location. In such cases, the structures are typically positioned and oriented by comparison to reference points or lines and are adjusted accordingly during installation. Aligning an extensive number of elements to a high degree of accuracy can also be challenging. For example, conveyor belts are assembled section by section and the whole line of the assembled sections must be aligned accurately. The accumulation of what might seem to be minor positioning errors for each individual section can result in unacceptable alignment of the overall system. In such a case, each element must be checked with adjustments made toward correctly positioning the entire assembled system. A cycle of positioning, checking, adjusting, and rechecking is executed until the complete system is installed correctly. This is indeed a time-consuming process. Moreover, some human factors research-studies in task-switching (Jersild, 1927; Allport *et al.*, 1994; Rogers and Monsell, 1995) have shown that the speed and/or accuracy is reduced when performing repeated sequences of different tasks compared to situations where the same single task is done repeatedly. This indicates that switching between the tasks of adjusting the location of the element and measuring it may limit performance. The stereo view of the 3D configuration of a structure or element of the structure, which is rendered onto the site in actual scale, location and orientation, may free workers from using measuring devices to check repeatedly the location of the structure or element in installation. As illustrated in Figure 3, workers would simply need to align a structure or element of the structure with its virtual configuration.

Figure 3. A conceptual image of installation task with AR aid.

3.1.4. Inspection

For inspection of the location of a critical structure or element, a survey team needs to accurately set up a measuring device, perhaps a theodolite or total station, at a reference and then take measurements and angles. If there are many critical structures and/or elements that require inspecting, it takes time to inspect them with such conventional inspection methods. Moreover, a survey team using conventional methods may have no way of recognising when reference points are incorrectly located due to prior measuring errors. However, display of the 3D configuration of the critical structures or elements rendered, in actual scale, location and orientation, onto the real scene of the corresponding installed elements, as illustrated in Figure 4, may free a survey team from

Figure 4. A conceptual image of inspection task with an AR aid.

needing to measure the location of the structures or elements with devices that require complex set up procedures. Using an AR-based method instead may save inspection time and prevent inspection errors since each object in the digital model is uniquely referenced to the coordinate system, thus lacking errors accumulated from intermediate references.

3.2. COORDINATION

A major potential application area for the use of AR in construction is to provide visual aids for coordination of the construction process. In this application, AR can be regarded as a medium for simulator output.

In coordinating the construction process, verbal descriptions, notes, or hand sketches for the present condition of work areas are conveyed to organise and determine upcoming work flows or resource allocation. However, such forms of communication are usually not complete, so field staff may misunderstand the work status and conditions. It may require a significant mental workload to correctly visualise comparison of the present condition or status of the work area with the 3D design and mentally simulate the paths of materials, equipment and/or workers based on this mental image. The significant mental workload, from a human factors' standpoint, hinders the field staff from recognising some conflicts between work procedures. Behzadan and Kamat (2006) presented the animation of construction activities superimposed onto a construction site. Their analysis indicates that AR can be used for construction simulation. However, considering the fact that coordination usually occurs in the office, multiple still shots of an AR scene may be more appropriate for coordination. Multiple photo views of a work area with a 3D design superimposed on it in correct scale and location may allow the field staff to understand the present condition of the work areas easily and free them from having to mentally superimpose such conditions on an image of the 3D design. The additional mental workload for simulating the paths of material, equipment and/or workers may be also significantly reduced by allowing the field staff to see and rearrange the computer-generated objects representing them inserted into the photo scene of the work area (Figure 5). This use of AR would essentially take interactive virtual simulation elements and place them in the real world context for the coordinator.

3.3. INTERPRETATION AND COMMUNICATION

Another key potential application area for the use of AR in construction is to provide visual aids for interpreting drawings and specifications and for communication. Such applications of AR can be regarded as enabling augmented CAD drawings and specifications.

Figure 5. A conceptual view of AR overlay for coordination.

Supervisors refer to design drawings and specifications to scrutinise whether or not work is performed as planned. Foremen also plan detailed procedures for a specific construction activity based on design drawings and specification requirements. Design drawings and specifications are also referred to for communication. For example, inspectors or field personnel convey some comments about a task to field personnel or foremen through a face-to-face communication to ensure the comments are understood. This process is typically accompanied by supporting nonverbal gestures (for example, pointing), observing work areas and presenting or describing relevant drawings and specification requirements.

During such exchanges, individuals infer a 3D mental image of the construction activity work process in terms of necessary tasks and resources, typically based on 2D design drawings and specification requirements. Envisioning the design in actual scale, position, and orientation from 2D drawings as it should appear in a real 3D space involves mental rotation and mental size transformation of the various views in the 2D drawings. Several research studies in mental rotation (Cooper and Podgorny, 1976; Shepard and Metzler 1971) showed that more time is needed to mentally match a rotated object with an original object as angular difference in orientation increases. Some research studies in mental size transformation (Bundesen and Larsen, 1975; Cave and Kosslyn, 1989; Larsen, 1985) showed that it also takes more time to mentally match a pair of objects that have the same shape but different size as the size ratio of the objects increases. These studies indicate that it may require a significant mental load to picture, in its intended real world location, the layout that would correspond to those depicted in conventional 2D construction drawings. A significant mental load may be also required to infer the three-dimensional design from the mentally laid out 2D drawings. Issa *et al.* (2003) showed that 3D drawings are preferred to 2D drawings as construction design becomes more complex. Their conclusion implies that the mental load experienced in inferring 3D design from 2D drawings increases as design complexity increases. For complex designs, even experienced participants may find it difficult to infer the 3D design from 2D drawings.

To reconcile written specifications with the drawings, construction field managers and supervisors must search through paper-based specifications and memorise requirements to mentally associate (match) them to the corresponding design drawings. Loftus *et al.* (1979) showed that short-term memory performance worsens as more information is retained. This finding indicate that field personnel may have more difficulty in both mentally coordinating the extrapolation of the design into 3D from the 2D drawings and in selecting critical annotations drawn from specification requirements when the designs of associated structures or elements are more varied and complex.

These factors point to the usefulness of a technology such as AR to reduce the cognitive load associated with interpreting abstract design representations. By AR, a selected portion of a 3D design model may be rendered spatially onto a field personnel's view of the construction site in actual scale, location and orientation, with virtual annotations of contextual data from the specification requirements (Figure 6). The less abstract model presented in full scale, with

Figure 6. A conceptual image of AR overlay of 3D design and contextual data.

critical details attached, may reduce significantly the mental work load for the field supervisor (or other responsible field personnel). It can enable the supervisor to see a meaningful representation of a structure, equipment, or component and associated contextual specifications information presented strategically to obviate the need to infer or search conventional design and specification documents often available only on paper.

So far we have explored AR for the construction phase as a medium for enabling more effective comprehension of the design by those involved in performing and monitoring construction. Now we turn our attention to critical technology considerations that have come to light through our efforts to develop prototype systems.

4. AR System Technology Issues

As Azuma *et al.* (2001) mentioned, enabling technologies for Augmented Reality include displays, tracking, and calibration. The same three enablers are still key today for creating any viable system for extensive use in construction. It is the characteristics of construction tasks and the construction site that dictate the necessity of resolving challenges in these three areas. The following sections briefly speak to each.

4.1. DISPLAYS

Head-mounted displays (HMDs) seem appropriate for AR in construction because HMDs enable workers to keep their hands available for typical construction tasks, such as assembling, installation, erecting, lifting, carrying, et cetera. However, HMDs are still somewhat weighty and bulky and typically tethered by video cabling. These features of HMDs restrict the usability and mobility of HMDs for construction practice. Although cabling problems will likely be minimised by using efficient data handling techniques and exploiting broadband wireless technologies such as Bluetooth, reducing the weight and size of HMDs is still a needed advancement. Beyond the technology developments that are wanting, HMDs also cause eye fatigue mostly due to images displaying close to the user's eyes. This effect is a serious human factor barrier to ultimate acceptance. User tests should explore use strategies that allow the user's eyes to rest from the strain of the short viewing distance.

HMDs are generally grouped into optical see-through and video see-through. Optical see-through HMDs provide a full field of view of the real environment, thus enabling workers to have more natural views of work areas. However, virtual images shown on optical see-through HMDs are not clear because they are not opaquely superimposed onto the real image scene. This effect might inhibit attention to important details in the digital content that is displayed. Also, the calibration of an optical see-through display for each user is difficult and time consuming. Optical see-through HMDs have to be recalibrated whenever they are taken off and put on as well as after each relative movement between the optical display and the user's eyes. Meanwhile, video see-through HMDs show virtual images clearly and once video see-through displays are calibrated, taking off and putting on displays does not influence the accuracy of the augmentation. However, they provide a limited field of view of the environment. In addition, a field of view of the camera that may be different from that of eyes can cause a distorted sense of space which might negatively influence the worker's ability to correctly assess the proximity of job site hazards. Based on these considerations, we conclude that HMDs have a reasonable potential to be utilised in AR in construction. However, there are also many technical issues to be addressed before HMDs can be embraced and used effectively on the construction site.

Although HMDs are attractive for supporting tasks that have hands-free demands, there are potential uses for handheld or stationary displays, particularly for layout and inspection. For example, the motion tracking technology developer, InterSense (2002) has demonstrated a mobile augmented maintenance application that incorporates a handheld computer monitor as the display device. It would be straightforward to apply the same type of technology to inspection tasks in construction where position is not the critical criterion. Shin (2007) made use of a tripod-mounted display to inspect the positions of steel columns. By using the stationary display, the dynamic error in this high-accuracy task was minimised.

4.2. TRACKING

Construction sites are characterised as expansive in nature. This indicates that tracking a system for AR in construction must be designed to cover a large range while maintaining fine accuracy. Even though there are some commercial tracking systems (for example, HiBall Tracker from 3rdTech and IS-1200 from Intersense) for limited indoor environments, which achieve high accuracy and several studies for large scale trackers (Azuma *et al.*, 1999; Behringer, 1999; Thomas *et al.*, 2000; You *et al.*, 1999), no tracking systems in research or on the market have satisfied the unique demands of construction sites.

Dunston *et al.* (2007) pointed out that a practical application of AR in construction must be mobile and might make use of predefined references since project site benchmarks, the project design, and in-place construction can be used as reference data as the user makes his or her way around the project site and through the constructed facility. Achieving the capability to facilitate performance in this regard would produce a system that might have broad application for the construction site.

Although we can readily identify options for predefined references for the user to chart his or her way around the project site and through the constructed facility, we are still challenged by the changing shape of the environment. In particular, the most challenging case of exploring 'uncharted territory' is not the one presented by the open building construction site. Navigating between and within operational spaces is the more challenging and worthwhile objective. It would be desirable, for example, for the tracking system to reveal where a pipe goes through the wall of one room and then, after transiting from that room to an adjacent room, automatically calculate and display where the same pipe enters through the wall. The real challenge then becomes one of establishing the combination of sensing technologies and recognition algorithms that provide reliable performance. These considerations are the drivers for determining the choice of tracking technology that will inspire confidence in AR applications.

4.3. CALIBRATION

As well as accurate tracking, accurate camera calibration methods are key to achieve accurate registration in AR systems. Even though some highly accu rate tracker methodologies are available, they are sensitive and inaccurate calibration can produce significant misalignment in registration. Many studies in AR have been performed to explore compelling calibration methods. To develop compelling trackers and calibration methods for AR systems for the construction site, the characteristic of the construction site should also be considered. Different elements of a constructed facility have different accuracy requirements with regard to position and orientation based on standards and technologically driven tolerances. Most of the studies in large scale AR systems have focused on developing accurate large scale trackers, but they have not considered the system accuracy as it depends upon the view distance. Shin *et al.* (2007) pointed out that unlike optically based AR systems, the accuracy of video-based AR systems depends on the view distance even with perfect calibration of the camera(s). For video-based AR systems on construction sites, this issue must also be addressed with the aim of achieving responsiveness to varied accuracy requirements.

5. Summary

This discussion of AR as a medium accessing information to support site operations and tasks reveals that AR can benefit three primary categories of activities that occur routinely in the construction phase: (1) building and inspecting, (2) coordination, and (3) interpretation and communication. In these application areas, AR may enable field staff or workers to perform their tasks more easily and effectively by providing necessary visual information for the tasks in a more convenient manner. However, technical issues such as displays, trackers and calibration methods must be addressed before AR can be used prevalently at the construction site. These technologies should be developed with critical consideration of the characteristics of construction sites if successful adoption of AR is to occur.

DISTRIBUTED AUGMENTED REALITY FOR VISUALISING COLLABORATIVE CONSTRUCTION TASKS

AMIN HAMMAD
Concordia University, Canada

Abstract. In this paper, a new methodology called Distributed Augmented Reality for Visualising Collaborative Construction Tasks (DARCC) is proposed. Using this methodology, virtual models of construction equipment can be operated and viewed by several operators to interactively simulate construction activities on the construction site in augmented reality mode. The chapter investigates the design issues of DARCC including tracking and registration, object modelling, engineering constraints, and interaction and communication methods. The DARCC methodology is implemented in a prototype system and tested through a case study of a bridge deck rehabilitation project.

Keywords. Distributed Augmented Reality, Tracking, Registration, Collaborative Construction.

1. Introduction

Augmented reality (AR) is a visualisation method in which virtual objects are aligned with the real world and the viewer can interact with the virtual objects in real time. Using a variety of 3D modelling, tracking, user interaction, rendering and display techniques, AR enhances users' immersion by allowing them to view the AR environment while moving in the real world. AR systems are classified as indoor or outdoor systems. Indoor AR provides an environment prepared with tracking infrastructure, limited user mobility, and stable lighting, but it limits users' actions, and therefore, it limits the types of AR applications (Azuma, 1997). On the other hand, outdoor AR should work in an unprepared environment.

In this research, we propose a new methodology for outdoor AR called Distributed Augmented Reality for Visualising Collaborative Construction Tasks (DARCC). DARCC provides a new approach for visualising multi-user collaborative construction simulation. The following scenario is used to explain

X. Wang and M.A. Schnabel (eds.), Mixed Reality in Architecture, Design and Construction, 171–183.

the vision of DARCC. Two or more users, equipped with head-mounted displays (HMDs) and tracking devices, interactively select, locate, and operate virtual construction equipment, such as cranes, on a construction site in order to check spatial and engineering constraints and rehearse for the real construction work. Each user can control one piece of equipment using joysticks from a third-person view in a way similar to real operations. If the actions of users do not violate the engineering constraints of the equipment (for example, those imposed by the loading charts of cranes), the virtual model of the equipment will react to those actions and the results will be visualised on their HMDs, augmenting the real scene. In addition, the information representing the state of the equipment is wirelessly transmitted to all other users and applied on clones of the equipment model so that each user can see the results of the actions taken by all other users on their respective virtual equipment. As a result, the team of users can share the AR scene resulting from their collaborative actions on the set of virtual equipment and can operate the equipment as if it exists on the construction site.

In order to realise the envisaged scenario, several issues of mobile AR need to be considered, such as accurate tracking and registration of real and virtual objects, user interaction with the system and with other users in real time, et cetera (Azuma *et al.*, 2001). In addition, suitable methods should be used to represent the virtual models and the engineering constraints related to the equipment. Furthermore, it is necessary to carefully design and implement the integration of a variety of technologies used in DARCC. The objectives of this chapter are: (1) To investigate the design issues of DARCC, including tracking and registration, object modelling and engineering constraints, and interaction and communication methods; (2) To investigate a procedure for applying DARCC in construction projects; and (3) To test the feasibility of the design by implementing and testing a prototype system.

2. Related Research

Tracking and registration are among the most important technological challenges of AR. Accurately tracking the user's position and viewing orientation is crucial for AR applications (Azuma *et al.*, 2001). Mobile outdoor AR brings more difficulties for accurate tracking because the outdoor environment is usually unprepared with the tracking infrastructure. However, tracking technology in the outdoor environment has been improving steadily in recent years (Azuma *et al.*, 2001).

Development in simulation software is making it possible to train crane operators to use computer graphics (Simlog, 2008) and to visualise the results of construction simulation (Kamat and Martinez, 2001, 2005). In the area of construction equipment research, Al-Hussein *et al.* (2005) developed a system, which can assist in selecting and locating cranes on construction sites using the information of load charts and working range. This system uses a crane database, named D-Crane, which has load charts of different manufacturers and the key dimensions of each crane including its carrier, main boom, jibs/ extensions, and accessories. The system ensures that the selected crane has the required lift capacity and can fit on site by satisfying a set of constraints described with detailed equations. However, the system only supports a set of predefined configurations of spatial constraints.

In the area of civil engineering, Webster *et al.* (1996) were the first to demonstrate the wide range of applications of AR in construction and inspection. Hammad *et al.* (2002, 2004) discussed the benefits of using AR and location-based computing to assist bridge inspectors in achieving their field tasks. Dunston *et al.* (2002) discussed the benefits of AR for design perception. Wang and Dunston (2006) studied the potential of augmented reality as an assistant viewer for computer-aided drawing. Behzadan and Kamat (2006) presented an application of outdoor AR for graphical simulation of construction activities. Kamat and El-Tawil (2007) used AR for rapid assessment of earthquake-induced building damage.

3. Proposed Approach

Figure 1 shows an overview of the design space of DARCC and identifies its main elements. These elements are: tracking and registration, dynamic object modelling and constraints, user interaction, and communication. The outdoor tracking is realised by combining high accuracy position tracking and a

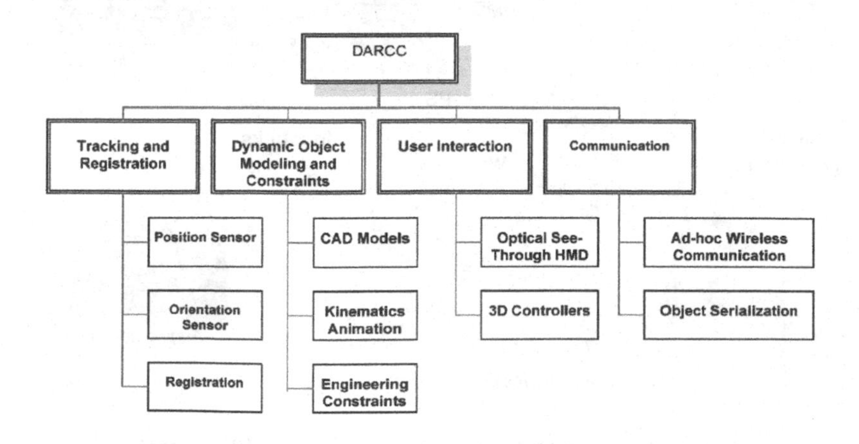

Figure 1. Design overview of DARCC.

three-dimensional (3D) motion sensor for orientation tracking. The models of the construction equipment are created as computer-aided design (CAD) models in a way that permits moving parts of the equipment (that is, kinematics animation) according to user actions while checking engineering constraints. If a constraint is violated by an action, the resulting movement is disallowed and a warning message is shown to the user. The user interaction is facilitated by an HMD as the output device, and 3D controllers (joysticks) as the input devices. The communication between users is realised using ad hoc wireless communication and object serialisation.

Figure 2 shows the conceptual configuration of the proposed system based on DARCC. A 3D orientation sensor provides the yaw, pitch, and roll angles which define the orientation of the user, and a GPS receiver provides the current position of the user in the world coordinate system. In addition, the wireless communication unit exchanges information with other users about equipment states.

The operation information input from the joysticks is sent to the scene generator. The scene generator accurately aligns and renders the virtual objects based on the information received from the above devices. Finally, an optical see-through HMD overlays the virtual scene from the scene generator and the real world scene as viewed with the HMD.

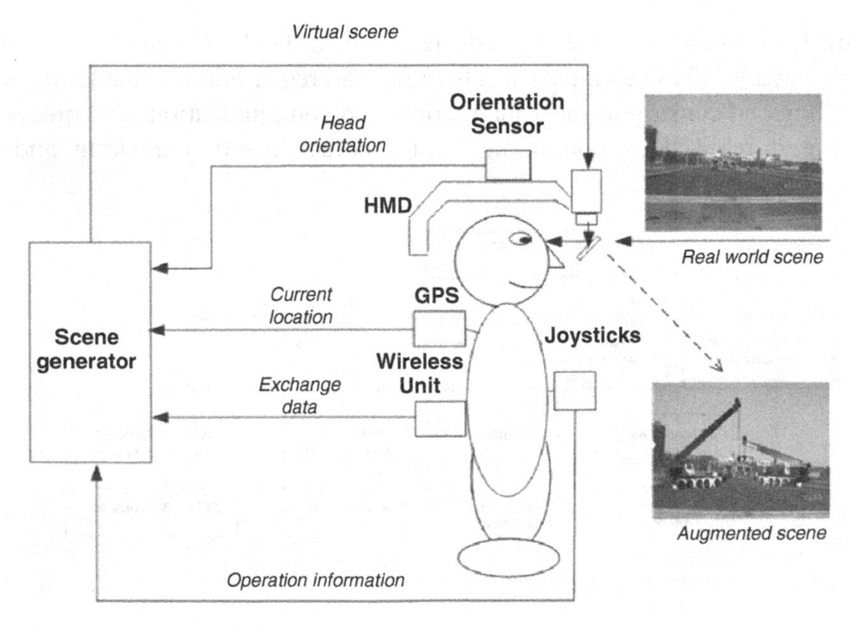

Figure 2. Conceptual configuration of the DMR system.

4. Tracking and Registration

Figure 3 shows the procedure of tracking and registration including two tracking parts: position and orientation tracking. The workflow is described as follow: (1) The computing unit calculates the position and orientation of the user's viewpoint relative to the world coordinate system based on the tracking results; (2) The computing unit calculates the position and orientation of virtual 3D objects in the world coordinate system; (3) The rendering unit renders the virtual 3D objects in one frame; and (4) The resulting frame is sent to the user's HMD. This process is repeated continuously to generate a stream of frames, giving the illusion of augmented reality.

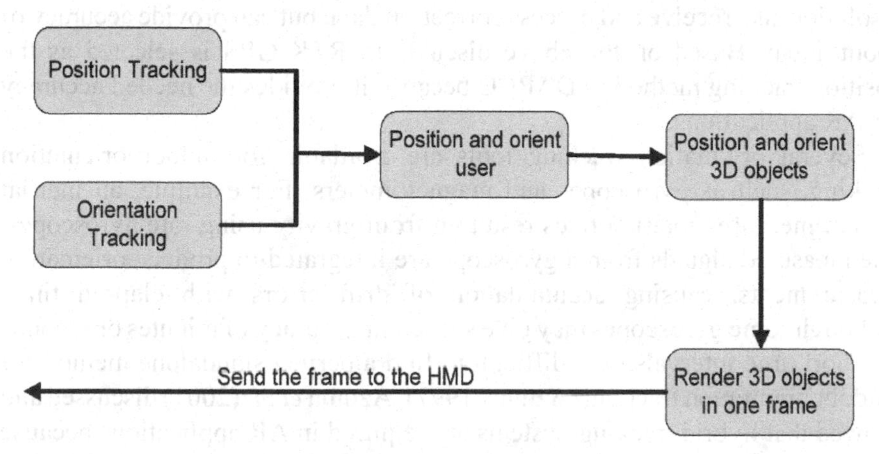

Figure 3. Procedure of tracking and registration.

4.1. POSITION AND ORIENTATION TRACKING

In order to select suitable tracking technologies, it is important to understand the principles of these technologies, and related accuracy and availability issues. The GPS is a positioning technology which is available anywhere with certain conditions and it measures the horizontal and vertical positions of the receiver from the GPS satellites. The GPS consists of 24 earth-orbiting satellites so that it can guarantee that there are at least four of them above the horizon for any point on global space at any time. The signals received by having a direct line of sight between the receiver and the satellites are used to find the position of the receiver using triangulation techniques. The factors that affect GPS accuracy include ionospheric and tropospheric distortion of the radio signals from the satellites, orbital alignment and clock errors of the satellites, and signal multi-path errors (reflections and bouncing of the signal near buildings). In addition, GPS is easily blocked in urban areas, near hills, or under highway bridges. The accuracy of a position is also a function of the geometry of the GPS constellation visible at that moment in time, i.e., when visible satellites are well separated in the sky, GPS receivers compute positions more accurately (Karimi *et al.*, 2004). One method to increase the accuracy of GPS is by using Differential GPS (DGPS). DGPS is based on correcting the effects of the pseudo-range errors caused by the ionosphere, troposphere, and satellite orbital and clock errors, by placing a GPS receiver at a precisely known location. The pseudo-range errors are considered common to all GPS receivers within some range. DGPS has a typical 3D accuracy of better than three metres, which is below the required accuracy in AR applications. Real-time kinematics GPS (RTK-GPS) receivers with carrier-phase ambiguity resolution also receive and process correction data, but can provide accuracy of about 1 cm. Based on the above discussion; RTK-GPS is selected as the position tracking method in DARCC because it provides the needed accuracy for AR applications.

Several orientation tracking tools are available for indoor orientation tracking, such as gyroscopes and magnetometers. For example, an inertial system measures rotation rates resulting from gravity using rate gyroscopes. The measured signals from a gyroscope are integrated to produce orientation measurements, causing accumulation of drift errors with elapsed time. Although some gyroscopes may give sufficient accuracy of minutes or seconds for short time intervals, it is difficult to find a perfect standalone method for outdoor orientation tracking (Azuma, 1997). Azuma *et al.* (2001) discussed and showed that hybrid tracking-systems are required in AR applications because each technology has limitations that cannot otherwise be overcome. The orientation tracking in DARCC should satisfy the following requirements: (1) Suitable hybrid tracking with scalable-range capabilities for mobile applications; (2) Easy to wear tracking devices; and (3) High update rates and accuracy. Therefore, a hybrid orientation tracking method (integrating gyroscopes, magnetometers and accelerometers) is adopted for our implementation as will be explained further in Section 6.

4.2. REGISTRATION AND THE VIEW MODEL

The crucial issue of registration is how to accurately align virtual objects with real ones (Azuma, 1997). The view model ensures that the user's eye position in the real world corresponds to the viewpoint position in the virtual world. This is realised by positioning the viewpoint at the same location at the user's eye relative to a geo-referenced world coordinate system. Several transformations are applied on the virtual coordinates when the user moves in the real world. Figure 4 shows the five principal coordinate systems that are considered in DARCC: Geo-referenced world coordinate system, W: (X_w, Y_w, Z_w); eye or camera-centred, C: (X_c, Y_c, Z_c); orientation tracking device-centred, O: (X_i, Y_i, Z_i); position tracking device-centred, P: (X_v, Y_v, Z_v), and 2D display coordinates, U: (X_u, Y_u). The transformation from O and P to C is introduced to calibrate the tracking results considering that the tracking devices are not centred at the eye or camera viewpoint. However, for simplification, the three systems (O, P, and C) are considered identical in this paper. The transformation from W to C is

$$W \rightarrow C: \begin{bmatrix} X_c \\ Y_c \\ Z_c \\ 1 \end{bmatrix} = R_{wc} T_{wc} \begin{bmatrix} X_w \\ Y_w \\ Z_w \\ 1 \end{bmatrix} \tag{1}$$

The rotation matrix R_{wc} and the translation matrix T_{wc} characterise the eyes' or camera's orientation and position with respect to the world coordinate system, respectively. The transformation from W to U is

$$W \rightarrow U: \begin{bmatrix} X_u \\ Y_u \\ Z_u \\ 1 \end{bmatrix} = K R_{wc} T_{wc} \begin{bmatrix} X_w \\ Y_w \\ Z_w \\ 1 \end{bmatrix} \tag{2}$$

where the matrix K represents the intrinsic parameters of the camera used in video see-through HMD, such as the focal length of the camera and the horizontal and vertical pixel sizes on the imaging plane. Because this research uses an optical see-through HMD, the rest of this section will focus on obtaining the matrices R_{wc} and T_{wc} based on tracking information.

In this research, the X and Z axes of the world coordinate system W composing the horizontal plane are matched with the two Cartesian axes used in large scale urban geographic information systems (GIS) maps, while the Y axis of W is matched with the altitude. This decision is very important because it allows us to use widely available GIS maps for locating virtual objects on the construction site. Consequently, the position information of the virtual objects and the tracking information can be both geo-referenced against the same world coordinate system W. Based on this matching, the V_x and V_z components of the translation vector are calculated by applying the Modified Transverse Mercator (MTM) projection algorithm on the latitude and longitude values obtained from the GPS receiver because this projection is used in most urban areas in Canada. The V_y component of the translation vector is equal to the altitude value obtained from the GPS receiver. Then, the translation matrix T_{wc} can be formulated as

$$T_{wc}=\begin{pmatrix}1 & 0 & 0 & V_x\\ 0 & 1 & 0 & V_y\\ 0 & 0 & 1 & V_z\\ 0 & 0 & 0 & 1\end{pmatrix} \tag{3}$$

On the other hand, the orientation sensor outputs the Euler angles as the orientation tracking results. These angles are the pitch (α), yaw (β) and roll (γ) for counter-clockwise rotations about the X, Y, and Z axes, respectively. The rotation matrices are given by

$$R_z(\alpha)=\begin{pmatrix}\cos\alpha & -\sin\alpha & 0 & 0\\ \sin\alpha & \cos\alpha & 0 & 0\\ 0 & 0 & 1 & 0\\ 0 & 0 & 0 & 1\end{pmatrix} \tag{4}$$

$$R_y(\beta)=\begin{pmatrix}\cos\beta & 0 & \sin\beta & 0\\ 0 & 1 & 0 & 0\\ -\sin\beta & 0 & \cos\beta & 0\\ 0 & 0 & 0 & 1\end{pmatrix} \tag{5}$$

$$R_z(\gamma)=\begin{pmatrix}1 & 0 & 0 & 0\\ 0 & \cos\gamma & -\sin\gamma & 0\\ 0 & \sin\gamma & \cos\gamma & 0\\ 0 & 0 & 0 & 1\end{pmatrix} \tag{6}$$

A single rotation matrix can be formed by multiplying the above three matrices to obtain

$$R_{wc}=R_z(\gamma)R_y(\beta)R_x(\alpha) \tag{7}$$

It is important to note that R_{wc} performs the roll first, then the pitch, and finally the yaw. The calculations of the translation and rotation matrix are not fully synchronised because of the different update frequencies of the position and orientation tracking devices. Thus we apply the two rotation and translation matrices independently to change the current viewpoint.

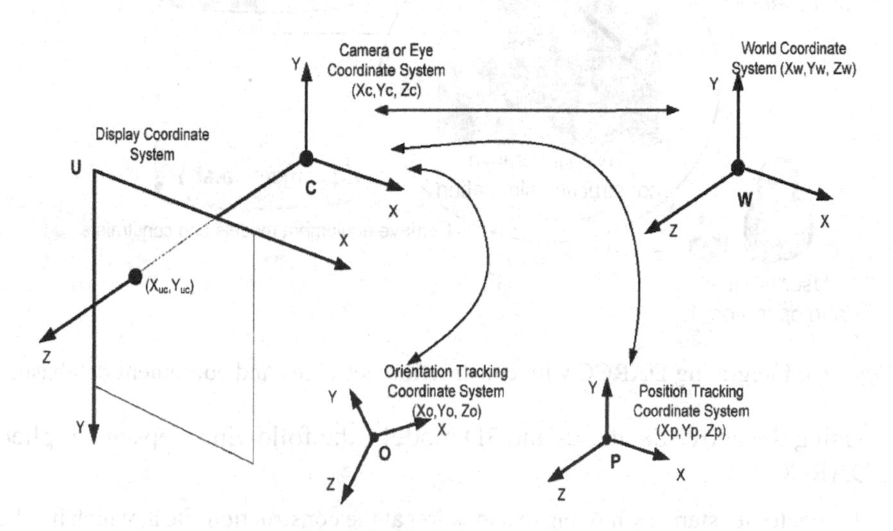

Figure 4. View model and related coordinate systems.

5. Procedure of Applying DARCC in Construction Projects

Figure 5 shows the concept of integrating DARCC with construction activities and equipment databases. The activity database includes information about all the activities in a construction project, such as the start and finish times of each activity, target physical components and their attributes, and types of equipment required in that activity. For example, in the bridge deck replacement project discussed in the case study, a typical activity is the replacement of an old section of the deck with a prefabricated panel. In this example, each activity will include the start and finish times, the ID number of the target section, and the required equipment such as cranes, trucks, etc.

The equipment database has the specifications about the different models of required types of construction equipment including the related constraints for using this equipment. Equipment manufacturers and large construction companies usually have databases of different equipment used in their work. D-Crane is a good example of such databases (Al-Hussein *et al.*, 2006). The equipment database used in DARCC extends available databases by adding equipment dynamic object models.

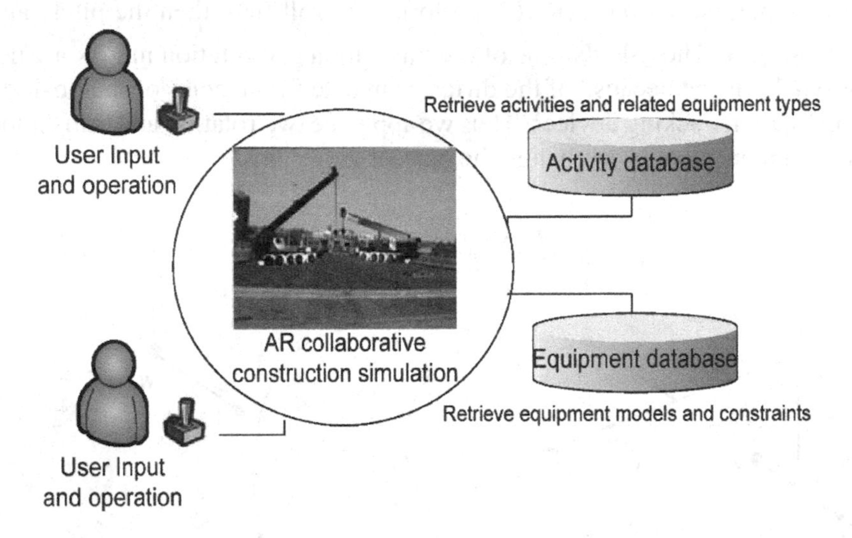

Figure 5. Integrating DARCC with construction activities and equipment databases.

Using the above databases and 3D models, the following steps are applied in DARCC:

1. Each user starts by moving to a location at the construction site at which he/she wants to observe construction activities from a third-person viewpoint.
2. The main activity to be considered in the construction simulation is selected by one of the users. The system retrieves the information about this activity and all other overlapping activities from the activity database. The information includes the related objects and the required equipment types.
3. Then, the equipment is selected by the users for each required type from the equipment database. It should be noted that selecting the optimal equipment is beyond the scope of this research.
4. The next step is to retrieve the equipment 3D models from the database. Other important parameters can be input by the users, such as the lift weight of a crane which is used in the simulation.
5. Locating virtual construction equipment and other objects on the construction site in DARCC is done interactively using a geographic information sub-system. The users click on a digital map of the construction site to approximately locate the virtual objects which are used in the selected activity. This digital map has the same MTM projection used in processing the position tracking for registration. The altitudes of virtual objects are extracted automatically from the digital terrain model of the site at the specified location.
6. In this step, the 3D models of the equipment are generated at the specified location on the construction site. The users interactively operate the equipment in the 3D augmented space of DARCC using the joysticks. They have the freedom to move around the construction site to observe the augmented scene from several viewpoints. The system continuously checks whether the manoeuvring operations are allowed by querying the engineering constraints.

6. Implementation and Case Study

A prototype system has been developed to test the proposed design. Table 1 shows the hardware devices used in the system for each user including an nVisor SX HMD display, two Saitek wireless Joysticks, a Panasonic Tough-Book tablet PC, a Trimble 5700 RTK GPS receiver, and an InertiaCube3 sensor.

Table 1. Specifications of devices used in the system.

Device	Type	Specifications
HMD	nVisor ST	Resolution: 1,280 × 1,024 pixels Vertical frequency: 60 Hz Brightness: 30 FL maximum (adjustable)
Joysticks	Saitek Cybord Evo wireless joystick	Range: 30 feet Frequency: 2.4 GHz
Tablet PC	Panasonic ToughBook CF-18	CPU: 1 GHz RAM: 512 MB
3D orientation sensor	InertiaCube3	Degrees of freedom: 3 (yaw, pitch and roll) Angular range: Full 360° (All axes) Maximum angular rate: 1,200° per second Minimum angular rate: 0° per second RMS accuracy: 1° in yaw, 0.25° in pitch & roll at 25°C RMS angular resolution: 0.03°

6.1. POSITION TRACKING

Trimble 5700 GPS receiver is used for position tracking. It is a 24-channel dual-frequency surveying level RTK-GPS receiver (Trimble, 2005). It can receive differential correction signals broadcasted from a stationary GPS receiver using a radio antenna or a Web-enabled cell phone in the case of Web broadcasting. The accuracy is about 1 cm under favourable conditions (direct line of sight to at least four satellites and availability of real-time correction data) and the update frequency is up to 10 Hz. The receiver is connected to the serial port of the computer or to a universal serial bus (USB) port using a COM-USB converter.

6.2. ORIENTATION TRACKING

The sensor InertiaCube3 (IC3) (InterSense, 2008) is used for orientation tracking. It provides a hybrid tracking solution integrating a gyroscope, an accelerometer, and a magnetometer along three perpendicular axes using Kalman filter techniques (Kalman, 1960). It measures nine physical properties, namely angular rates, linear accelerations, and magnetic field components along all three axes. The angular rates are used to compute the orientation of the sensor with low latency and power consumption in real time. The maximum and minimum angular rates are 1,200° and 0° per second, respectively. The IC3 sensor connects to the computer wirelessly using a 2.4 GHz radio module that allows up to 16 different channel selections. The radio module has a very low latency, low power consumption, high bandwidth and wide range. However, using the IC3 sensor or receiver near large metal surfaces would reduce the accuracy of the tracking because of the metal effects on the magnetometer. Moreover, because the IEEE 802.11 wireless communication devices of DARCC use the same frequency of IC3, the IC3 should be set to a channel different from these devices to avoid interferences.

In the case study, we used DARCC to simulate the deck rehabilitation project of Jacques Cartier Bridge in Montreal, Canada (Zaki and Mailhot, 2003). Two users operate two virtual telescopic cranes on the real bridge for installing a prefabricated panel. The users start by interactively adjusting the initial position and orientation of the virtual cranes using the wireless joysticks. Then they collaboratively operate the virtual cranes to achieve the construction activity. Figure 6 shows an AR scene of a test where the image of a construction site of the bridge is augmented with two virtual cranes. The GPS receiver and 3D orientation sensor track the user's position and orientation and the tracking information is used to apply the necessary transformation matrix on the viewpoint of the virtual scene. Therefore, the pose of the virtual cranes follow the user's movement in the augmented scene. The test represents the user interaction and tracking ability in outdoor environment. In addition, the engineering constraints have been successfully tested.

Figure 6. Two cranes replacing a bridge deck section with a prefabricated panel: (a) picture of the real project; (b) snapshot of AR simulation.

7. Conclusions

This paper proposed a new methodology for interactive construction simulation using outdoor augmented reality techniques. The methodology is called Distributed Augmented Reality for Visualising Collaborative Construction Tasks (DARCC). The contributions of this research are grouped into the following areas:

1. The design of DARCC identifies the requirements for selecting accurate tracking and interaction devices for AR-based construction simulation. It also integrates novel methods for dynamic object modelling, engineering constraints processing, and multi-user real-time communication in the outdoor environment.
2. The registration method of DARCC is new in that it adopts the coordinate system used in GIS projection as the base for the geo-referenced world coordinate system of the AR registration. This method facilitates the mapping of real world objects and virtual objects based on widely available digital maps.
3. A procedure of applying DARCC in construction projects has been developed. This procedure can be easily applied in practice because it is based on available information of construction activities and equipment.
4. A prototype system was developed in Java implementing the design of DARCC. This system integrates state of the art tracking devices for accurate tracking, including an RTK-GPS receiver for position tracking and a hybrid 3D orientation sensor. A high-end optical see-through HMD and wireless joysticks are integrated in the system for user-friendly interaction. Several software problems were solved to integrate the hardware devices while keeping the system modular and extensible. The favourable initial testing results of the system showed that it has potential for practical applications.

Acknowledgments

The author would like to thank the contributions of Mr. Hui Wang in developing the system introduced in Section 4 and in preparing the text.

AUGMENTED 3D ARROWS REACH THEIR LIMITS IN AUTOMOTIVE ENVIRONMENTS

Why Are AR Schemes Confusing?

MARCUS TÖNNIS AND GUDRUN KLINKER
Technische Universität München, Germany

Abstract. 3D arrows are a widely accepted Augmented Reality (AR) presentation scheme for various applications, ranging from picking tasks over indicators of directional warnings, to navigational aids. Applying AR schemes to the automobile sector, where arrows are already in use for navigational tasks requires developers of AR-based systems to think beyond arrow-based visualisation. Further presentation schemes require either easy differentiation between the semantic contexts of virtual arrows or should preferably use other shapes to transfer their message. Our group has investigated AR-based advanced driver assistance systems for several years now. We began our research by investigating certain issues of the automotive domain; we have built and evaluated separate solutions, sometimes using arrow schemes. To cope with the increasing amount of in-car systems and their user interfaces we now attempt to incorporate all approaches into a single integrated human-centred system. Here we illustrate our separate applications, collaborative design and evaluation platforms, then come to the issue of an integrated solution. From our work we summarise experiences which can facilitate design processes for AR applications in general.

Keywords. 3D Arrows, Car Driver Assistance, Human-Centred Systems, Human Factors.

1. Introduction

Spatially related information is one of the main application domains of Augmented Reality (AR). The three dimensional world becomes enriched with virtual objects designed to generate comprehension of spatial relationships or to support certain activities by showing how and where to apply an action. Main components for the end user of AR are presentation schemes, mostly using the

X. Wang and M.A. Schnabel (eds.), Mixed Reality in Architecture, Design and Construction, 185–202.

visual channel. Such presentation schemes refer to real objects directly, indicating them or showing further information about them. Other applications of AR-based presentation schemes refer to navigation tasks, where an object or location is not indicated directly but the route is enhanced through visual presentation schemes.

As a matter of fact, arrows, and specifically 3D arrows, are often used to guide to or indicate places and objects in the user's environment. The shape allows for a wide range of applications ranging from picking tasks in the logistics domain over indicators of directional warnings in time-critical environments, to navigational aids for way-finding tasks. Arrow-based schemes enable a variety of transmittable semantic context.

Research conducted in our group often incorporates arrow-based schemes as well. One area of investigation focuses on the combination of Augmented Reality and driver assistance systems. Arrows are already a well known shape in the automotive domain. On signs, they indicate roads the driver is allowed to take or give hints for the route to the next city or point of interest. Arrows on the ground indicate lanes and their heading, especially at crossings.

The first approaches to develop novel AR-based presentation schemes for car driver assistance let us neglect the existence of the large number of arrows used in the automotive environments. We did not completely expel that knowledge from our minds, but wanted to think freely to generate high performance solutions for problems we experienced as car drivers. Later, as we had a number of AR-based solutions for spatially related issues, we began evaluating how we could integrate these separate into one user-centred system. This approach is necessary, because the automotive environment is time-critical in a variety of ways. Assistance systems must not only be interruptible at any point in time so that the driver can refocus his attention back to the road and the traffic, but it must not require a huge amount of the driver's mental workload. They must be interpretable and projectable according to their spatial dependencies into the surrounding environment, which can quickly change, requiring the driver's cognitive processes to focus on issues of manoeuvring and stabilisation. The highest priority of assistance systems is to support the driver in achieving optimal performance under any kind of traffic condition. Secondly, such assistance systems must refrain from giving too much information to the driver at once. Information overload increases the risk of incorrect interpretation or prioritisation and separated systems can contribute to driver irritation through un-harmonised user interfaces.

We have investigated a driver's workplace under the paradigm of AR, based on the concept that 3D relationships are transmitted best in 3D. The results are several AR-based solutions for advanced driver assistance systems. In this chapter, we first investigate a driver's workplace, reveal issues for car drivers and illustrate issues for (visual) assistance systems of the next generation. Then we illustrate some of the separate applications that we have built to support drivers in their task of driving. In order to explain how these systems have been developed, we illustrate the design processes and experiences gained from studies in a driving simulator. Later, we discuss the issues related to an integrated human-centred solution with a harmonised user interface. Summarised approaches are investigated to facilitate design processes for AR applications in general.

2. Advanced Driver Assistance Systems

To manoeuvre safely, a car driver on the road has to react to many stimuli from the surrounding environment and from inside the car. All these stimuli are necessary input to control the car correctly, to keep it on the road and not to hit any obstacles. Other stimuli are caused through less important outside objects and events, or they come from extra, tertiary systems in the car, for example, to enhance user comfort.

Modern cars implement various different systems as a means to differentiate themselves from competitors and to co-exist on the market. All such in-car systems can be categorised based on their goals towards comfort and safety. One can distinguish between driver In-Vehicle Information Systems (IVIS) and Advanced Driver Assistance Systems (ADAS). Driver information systems are designed to inform and entertain the user while driving. Examples are air conditioning control, radio control and car navigation systems. ADAS systems, on the other hand, support drivers with their driving task. Examples are heading control or distance control (e.g. ACC – maintains speed in respect to a leading car). They support drivers in environmental surveying and they inform the driver about insecure manoeuvring activities or overtake certain responsibilities of the driving task. Thus ADAS systems are active components for increasing safety.

2.1. DRIVER'S WORKPLACE

To investigate how AR-based driver assistance concepts can facilitate safety aspects, the driver's workplace requires some classification to determine promising application areas. Car drivers perform interactive and concurrent activities that can be divided into three classes: primary, secondary and tertiary tasks (Geiser, 1985). Primary tasks describe how to manoeuvre the vehicle itself. The driver controls the heading and speed of the car, as well as the distance to other cars or objects. Secondary tasks are mandatory functions such as setting turning signals or activating the windshield wiper. They increase safety for the driver, the car and the environment. Tertiary tasks cover entertainment and information functionality. These do not have any direct relationship to the driving task. Rather, they provide luxury services which are in high demand from today's car buyers, and thus a mandatory asset in modern cars.

The primary task of driving can again be split into three responsibilities of a driver: Navigation, Manoeuvring and Stabilisation. Figure 1 illustrates these subtasks in sketches. While driving, drivers always have to know about the route to their destination, and within the current immediate traffic environment, drivers have to manoeuvre safely, not to hit any obstacles; and finally they have to stabilise their cars, such that they do not run off the road or to inadvertently cross into other lanes.

Figure 1. Three responsibilities of a driver in the primary task: navigation, manoeuvring and stabilisation (adopted from Bubb 1993).

Management of these concurrent primary, secondary and tertiary driving tasks require a significant amount of human capability. In general, there is no problem for trained car drivers, but due to absent-mindedness or in critical situations a driver may neglect important events and can thus cause an accident.

This is the point where ADAS systems come into action. They monitor the car's environment and warn the driver about certain situations, events or obstacles. Ultimately, some can also take over control to prohibit or at least mitigate an accident. Other approaches of assistance systems concurrently provide safety related information in an ambient manner. An often-overlooked tool here is the speed indicator: constantly visible, it allows the driver to maintain legal speed or decrease speed correctly before a curve. Fully automated driving as a final goal is still a topic of research, but, for instance, first collision avoidance systems and systems capable of lane keeping and convoying already exist. Current state-of-the-art systems from major car manufacturers include, lateral assistance systems that warn drivers about unintended lane departures or longitudinal systems that adapt to a leading car's speed. Those systems only operate within certain safety margins and turn themselves off automatically if they can no longer maintain their task in an appropriate manner.

There are various approaches to transmit an assistance system's message to the driver. Aside from the active systems that directly affect the car's driving state, some systems use human sensory channels, such as auditory or tactile senses. The most used channel is the visual channel. Especially Head-Down

Displays in the instrument cluster and the central information display draw a driver's attention off the road to get their message. A major contribution of AR can come through the relatively newly introduced Head-Up Display (HUD). It already reduces off-road glancing times due to reduced focal adaptation time and smaller saccade angles.

Large scale HUDs can superimpose spatially related information and enable a new field of concurrent information visualisation. Heavily transmittable values of the driving state, such as braking distance or the direction towards safety critical objects become displayable and accessible in a new way.

2.2. VISUAL AND COGNITIVE ISSUES OF DRIVER ASSISTANCE SYSTEMS

The importance of HUDs in cars will grow significantly, as soon as it is technically feasible to project large quantities of information in high resolution onto the windshield. Eye-tracking technology will then enable Augmented Reality (AR) to embed dynamic presentation schemes into the environment, thereby minimising glance distraction time. For instance, when virtual navigational arrows are placed onto the road in front of the car, drivers do not have to look at the car-internal navigation display.

This upcoming technology deepens questions and problems pertaining to information overload, perceptual tunnelling and cognitive capture (Tönnis *et al.*, 2006):

- **Information Overload** refers to the state of having too much information to make a decision or remain informed about a topic. Large amounts of currently available information, a high rate of new information being added, contradictions in available information, a low signal-to-noise ratio, and inefficient methods for comparing and processing different kinds of information can all contribute to this effect.
- **Perceptual tunnelling** is a phenomenon that originally comes from aviation and in which an individual becomes focused on one stimulus, like a flashing warning signal and neglects to attend to other important tasks/information such as driving the car.
- **Cognitive capture** refers to the situation where the driver may be totally 'lost in thought,' a condition which, in particular, could impair situational awareness. Where emotional content (that is, personal involvement) in a conversation is high, such as arguing with someone over the phone, the likelihood of cognitive capture is increased. Instruments that require some level of cognitive involvement can lead to a loss of situational awareness and are viewed as increasing the risk of an accident.

When embedding additional visual information into the real world, information density is not computable, because the outside environment of the car is changing during travel. If many events that are relevant to driving occur at the same location on the windshield and/or at the same point in time, they can generate one or more of the announced problems.

3. Augmented Reality-Based Driver Assistance

Research into HUDs for car drivers enables visualisation of spatially related information and thus enables a new field of concurrent information visualisation. Assistance system visualisation concepts can set a focus on the presentation of continuous information if such visual stimuli do not lead to the mental factors described above. Such AR-based concepts furthermore can enable new approaches for warning schemes. With reduced recognition times preconditioned due to in-place visualisation, the informative content of warning schemes can be increased.

Here we illustrate some of our assistance system concepts that we have built and tested to assist drivers in their task of driving.

3.1. GUIDANCE OF ATTENTION

As driver inattention is a primary cause for up to 78% of crashes and 65% of near-crashes (Klauer *et al.*, 2005), driver support systems should help reduce driver inattention. To this end, an ADAS system needs to monitor and track the car's environment, exploiting the increasing availability of sensors to detect imminent dangerous situations in traffic and other obstacles in the car's near distance. Such sensor data must be provided via suitable output channels to catch a driver's attention and to guide him in that direction.

Dealing with the issue towards guiding a driver's attention without distracting them from the driving task, we have generated an AR-based presentation scheme (Tönnis and Klinker, 2006) and tested it in comparison to common other presentation schemes.

If an imminent danger occurs, for instance an upcoming crash, the driver not only has to be informed of the existence of such a danger, they also need to know where in the environment the dangerous situation is located. It is conceivable to use bird's eye view presentations schemes that are already used in parking assistance systems that provide drivers with distance information when they park their cars. Yet, in contrast to a parking scenario, dangerous situations can occur during driving at much higher speed. The mental load of an exocentric presentation may be too high and time consuming for drivers. They mentally have to transform their view into the exocentric bird's eye perspective, get an overview of spatial relationships in that perspective, transform them back and then determine the specified location with respect to their own egocentric position.

Using an egocentric AR perspective reduces the effort of transforming between several coordinate frames. Unfortunately, a dangerous situation may not be in the driver's field of view. Therefore a mere location indicator is not useful – the driver's attention needs to be guided. A compass metaphor appears to be more promising for use in cars.

Our AR-based scheme is based on a 3D arrow floating in front of the driver's head, pointing into the direction of the danger. The direction of a simple 3D arrow made out of a cone and a cylinder is difficult to interpret with respect to its direction when it is pointing directly forward or backward. Attaching fins on the rear of the arrow dissolves this ambiguity.

A further problem in identifying the exact orientation of the arrow comes from the fact that the driver cannot see the exact location of the floating arrow in front of the driver at the distance of the front bumper. Therefore some people mentally translate the arrow to their own head's position. As a result, the imminent danger is searched for in an orientation that is too far backward. To avoid any ambiguity about the exact location, we have attached the arrow to a vertical pole at the location of the 3D arrow. From the driver's perspective this pole seems to be mounted at the front of the car. The focal distance of the virtual image in the self built large scale HUD was placed at the same distance.

AR visualisations must not cover large areas in the windshield, because obstacles in upcoming traffic could be occluded. A minimal volume for indicating various directions of the arrow is reached by rotating the arrow around its midpoint rather than its back end. The arrow is mounted accordingly on the pole. Figure 2 shows the 3D arrow from the driver's point of view.

For enhanced attentional capture and increased spatial perception of the arrow's direction, a short animated rotation has been implemented. The 3D arrow appears in front of the driver, fixed to the pole. At this moment it is pointing in a direction that is 45 degrees off from the direction of the imminent danger, oriented forwards. The arrow immediately starts rotating horizontally

Figure 2. The AR-based guidance of attention scheme consists of a 3D arrow with three fins at the rear-side. The arrow is mounted on a pole and thus virtually attached to the front bumper.

by 45 degrees into the direction of the imminent danger. This animation takes 0.25 s. The rotation around the horizontal axis has been chosen, because it is the most familiar rotation for ground traffic. All other rotations would have to include vertical rotations and therefore require more spatial interpretation.

We have compared this presentation scheme to a bird's eye perspective scheme that is capable of displaying locations close to a car. We decided to use the compass metaphor here as well. Our second presentation scheme used a bird's eye sketch of the car with a 2D arrow pointing into the direction of the upcoming danger. The arrow has been placed in front of the car silhouette, Figure 3.

Figure 3. The bird's eye presentation scheme for guidance of attention. A 2D arrow is placed in front of the car with a slight overlap.

In principle, the bird's eye presentation scheme is almost a projected visualisation of the 3D arrow scheme, seen from above. All these design decisions attribute to the generation of a spatial relationship between the arrow and the car, without indexing the car itself.

We have set up a test environment in a fixed base driving simulator. It consists of a 50 degree field of view rendering rural road course. The test was performed with 24 test subjects. We measured the effective response time, average mistakes, error quotients, mean lane departure and speed variation. In advance, the test subjects filled out a subjective questionnaire.

The results (Tönnis and Klinker, 2006) show that guidance of a driver's attention towards the direction of such danger while remaining in the driver's frame of reference is superior to a presentation in another frame of reference. Primarily, detection times are significantly reduced. Other factors such as speed and lane deviation are not worse than with the exocentric warning scheme. In time-critical situations, the reaction time is the most important factor for safety. Thus, this is the most relevant achievement. Reducing the reaction time by concurrently reducing driver distraction is the main goal to achieve when testing spatial alerting systems.

3.2. VISUALISING A CAR'S PHYSICAL STATE

Statistics on accident counts reveal that many accidents are caused by human errors in longitudinal and lateral car control. Lateral accidents occur due to lane departure or collisions with lateral traffic in adjacent lanes, while longitudinal collisions occur due to obstacles, upcoming traffic or rear end collisions.

Drivers act in a tight control cycle, in which they continuously perceive the environment, interpret the current situation and execute the most suitable action to control the car. Assistance systems with secondary displays are not integrated into this tight control loop. While looking at and reacting to a warning signal on a secondary display, drivers are taken out of the loop (Brookhuis and de Waard, 2006).

In our investigations toward a general concept of safety for car drivers (Tönnis *et al.*, 2007), we focus on an integrated approach to keep drivers in the loop of the control circuit. It does not warn drivers about nearby critical situations, but rather shows them where they are going. Drivers can stay in the loop of the control circuit and continuously perceive the actual state of the car. Our concept incorporates a predictor for the pathway, intended to improve driving performance directly and does not wait until a certain critical event, like a lane departure, has arisen. Drivers are no longer required to pull their attention away from environmental settings to a secondary display. Rather, they can concentrate on surrounding traffic.

Taking into account that AR enables visual presentation schemes from an egocentric point of view, we use the in-car control devices to control augmented objects that are embedded into the personal view of the driver. Such visual presentation schemes in the Head-Up Display (HUD) of the car indicate how drivers are manoeuvring through the 3D environment, specifically, where their car is heading at the moment. We assume that this kind of assistance is much less distractive than secondary displays, because it keeps drivers in the loop of the control circuit. This concept also allows for future integration of safe distance indication in platooning traffic where the risk of rear-end collision is increased.

We have developed two HUD-based visualisation schemes that present a car's intrinsic status as a function of lateral and longitudinal properties of car motion. The first presentation scheme consists of a single bar shown in front of the car on the street. The second scheme extends the first one by outlining the path that will be covered by the car, Figure 4.

The bar in the first scheme indicates where the car would come to a halt, if the brakes were fully pressed at the current point in time. Depending on the steering angle, the bar turns left or right, according to the way the car will turn. The braking bar assistance scheme is a flat cube, 2 cm high, shown with the same width as the driver's own car. It is 50 cm long such that the bar is visible

Figure 4. The braking bar and the drive-path in a light left curve. One can see the car actually is slightly too far to the left and starts the left curve too early.

even at high speeds. Due to the thin layout, the bar does not occlude a large area in the driver's field of view. The bar is coloured in bright green, known to be well suited to the presentation in HUDs – where dark colours are not visible. The colour has good contrast to common grey scales on roads or unpaved brown country lanes. The bar is rendered in perspective such that its size becomes smaller, when speed increases and the braking distance thus becomes longer. Turning the steering wheel causes the bar to rotate and move left or right – according to the curved path the car will take at the current turning radius. The lateral placement of the bar is computed according to the driving model of a single track. The bar, in addition, rotates around the vertical axis, so that it shows the stopping line of the car's front at every point in time.

The second presentation scheme shows the car's drive-path. To support the estimation of curves and narrow road sections, we have explored the benefit of visualising the area through which the car will drive on its current trajectory. The drive-path indicator extends the braking bar presentation scheme by two additional sets of polygons that connect the bar to the car. Here the right and left border of the bar are connected by polygons to the right and left front corner of the car. Each of the two polygons uses four vertices between the car and the bar to generate a rounded shape. These lines surround the area, which will be covered by the car. The drive-path-based presentation is intended to better convey the alignment of the driving path with curves in the road.

Both visualisation schemes have been implemented and tested in a driving simulator against a baseline system without any assistance. We have used a fixed-base driving simulator with a 40 degree field of view. To reach the effect of no focal adaptation as a conformal HUD would have, the assistance scheme is incorporated into the rendering system of the rural scenery and projected onto the projection wall. Twenty-seven participants have been tested in our study. We recorded objective values of speed and lane deviation behaviour and conducted subjective questionnaires.

Results (Tönnis *et al.*, 2007) show that the test subjects drove faster with increasing visual assistance. Since driving simulators are safe environments, higher speeds could be expected, but the further increase shows that the visual aid again raises a feeling of safety. Another fact in longitudinal behaviour is found in the standard deviation of speed. The drive-path scheme oscillates significantly more than both other schemes. Here the drivers seem to neglect their obligation to maintain proper speeds when they look at the animated presentation scheme of the drive-path.

The lateral assistance appears useful for lane-keeping behaviour because the lane deviation decreases the higher the visual assistance is, but the drive-path scheme oscillates more than the pure bar scheme. Summarising subjective measurements, test subjects judge an improved overall driving quality for the bar scheme in comparison to no assistance, especially the findings that the bar scheme does not increase overall workload (NASA TLX, Hart and Staveland, 1988). It reduces lane deviation and does not increase oscillations in speed and lateral movement which make this scheme interesting for further analysis. The facts, that visual assistance brings an inaccurate feeling of safety and that common design principles for visual aids in time-critical systems enforce presentation schemes to be as minimal and easy to perceive as possible, let the drive-path scheme appear to be the most interesting candidate for further extension to a platooning aid. Thus, the bar scheme should be preferred over the drive-path scheme.

3.3. NAVIGATIONAL ARROWS

Just like real, physical arrows and other marks that are painted directly onto the road at intersections, navigational arrows could be augmented onto the road as well. This can be especially useful at crossings, where more than one road departs, or in other ambiguous situations. The implicit direct indication of which road to choose can be a great facilitation, especially when the driver does not have to look at a secondary navigation display nor need to interpret distance information of the navigation system. Figure 5 shows such an augmentation for a section of open road.

This augmentation is not ideal in cases where leading traffic causes occlusions, particularly when the traffic is dense, as the driver could not perceive the direction of the arrow until the directly leading car has left its position. An alternative for this could be to place the arrow in front of the occluding object, as shown in Figure 6. But even if the arrow is semi-transparent, it occludes a certain field of the environment, increasing the risk of not perceiving important changes in traffic. Furthermore, the depth cue perception of the

Figure 5. An augmented navigational arrow on open road.

Figure 6. A navigational arrow with reversed depth cue. Should the car occlude the arrow or not?

driver is reversed by such a visual presentation. Such a reversed depth-cue makes it more difficult for the driver to estimate where the arrow would belong, as related to its correct distance and depth to the driver. The 3D relationship is altered, so that the overall spatial relations are more difficult to interpret.

Alternatives to placing navigational symbols on the street are under investigation in many problem domains. For instance, Smith and Hart (2006) have evaluated various presentation schemes for way-finding tasks in virtual environments, measuring their cognitive loads. Among other schemes, a 3D graphical plan was evaluated. The 3D graphical plan was presented as a long ribbon floating some metres above the ground, ranging from the start to the destination. From all evaluated schemes, the 3D graphical plan required the highest mental load, probably due to an unfamiliar requirement for looking upwards.

We are currently investigating the concept of augmented navigational arrows. We plan to perform the evaluation in a real car with a built in large scale HUD that allows for location-fixed rendering of navigational arrows.

4. Issues Towards an Integrated Concept for Car Driver Assistance

Each separate application for AR in the automotive environment has also contributed to knowledge acquisition for the broader AR community. Intuitive understanding and usability experiences of our concepts have either been proven or look promising for further research. However, the combination of all aspects into future systems that use AR requires further investigation.

The investigation of integrated concepts is necessary in generating a user interface without ambiguous presentation schemes. Such an intuitive concept has to ensure that the user interface never provides too much information at any point in time. Only such an integrated human-centred solution is applicable in the automotive or any other time-critical environment, otherwise there could be too much information or information difficult to interpret and thus the risk of critical situations is increased again. The desired benefit of ADAS systems would disintegrate merely because of the characteristics of the system.

Different approaches are possible to investigate integrated systems. Tasks of the application domain can be analyzed to classify areas of application and frames of reference to mount or to relate presentation schemes to the user can be examined to find a suitable classification for the application of arrows.

4.1. TASKS OF MOBILITY

A suitable point to start investigations towards an integrated approach lies in the area of mobile applications. Here AR-based concepts in general have to deal with the same issues as in the automotive domain. General AR systems for mobile applications often deal directly with the aspects particular to mobility. Various aspects flow into such applications; primarily these are:

- Way-finding/travelling (manoeuvring): to know the route to the destination and to find the correct way
- Surveying: to check the environment for possible obstacles or interesting things/ information
- Stabilisation: to stay on the correct track

Research in the AR community mainly focuses on way-finding and travelling issues. Here approaches range from arrow-based compasses to superimposed pathways (Smith and Hart, 2006). 2D arrows are commonly used in other systems (Narzt *et al.*, 2003) to indicate certain turns. Reitmayr and Schmalstieg (2004) use waypoints and interconnecting lines.

To improve surveillance of the environment, superimpositions of objects (Lindl and Walchshäusl, 2006) are used as well as (again) arrow-based representations (Tönnis and Klinker, 2006).

Supporting stabilisation tasks with additional arrow-based presentation could use arrows to indicate the lane or road shoulder to be passed or visualise speed vectors. The immense amount of semantic content an arrow scheme can transmit would surely make all arrow-based schemes ambiguous when they need to be convey specific meaning. The bar and drive-path presentation scheme presented here contributes to stabilisation tasks as well, but uses a different visual representation.

To allow integrated systems that facilitate all three interaction tasks, the intuitiveness of symbols must be guaranteed. Especially for time-critical applications, e.g., surgery, catastrophe management or, as in our case, car driving, where AR can become a supportive system, the design of each assistance system should not interfere with one another or has to carry implicit knowledge regarding the meaning of every aspect of each metaphor.

As arrows are well known for way-finding and travel tasks in any kind of navigation system, independently of whether or not they are displayed on in-car-monitors or painted on pavements, arrows should remain the major presentation scheme to guide a user into a certain direction.

Issues of surveying can be distinguished in two further classes: objects that are in sight of the user and objects which are not in the field of view of the user. In the first case, direct object highlighting is the general approach. Over time this presentation scheme becomes familiar to any kind of user, as an increasing number of people are working with computers and understand object selection, while another increasing set of people know about highlighted objects from computer games.

The case when a dangerous approaching object is not in the personal field of view is more difficult to handle. AR is useful in superimposing information if it is visible, but it can easily be relegated to an aid for indicating direction. Here, spatially aligned 3D arrows have been examined and proved to work well (Tönnis and Klinker, 2006, see above). Among other things, these arrows were extended by spatial sound, making the impression of danger more imminent and thereby turning the indication metaphor into a warning scheme. The level of awareness is raised by using spatial sound for warning purposes which, in general, appear more rarely than navigational aids and are thus less annoying. This is built on the expectation that multi-modal schemes are more intuitively understood in situations where danger is outside the field of view and optical signals within the field of view must contain a level of indirection. They cannot be placed directly at the dangerous spot but tell the driver in what direction to look. Alternative approaches exist for 2D applications. Baudisch and Rosenholtz (2003) used circular segments in restricted viewports to indicate hidden objects nearby. Such indirect concepts require further investigation for their applicability in 3D environments.

A further important direction of research involves the development and evaluation of schemes to help users anticipate dynamic motion behaviour of other objects. For the automotive sector, we have designed a drive bar (Tönnis *et al.*, 2007). This approach is easily transferable to any kind of vehicle by adjusting the bar's width. Findings show that presentation schemes of this concept generate a feeling of safety while not having a negative impact on the hedonic qualities of the user interface. Results such as these makes this concept promising for research in other application areas. Furthermore, the results show that keeping trajectory and estimating speed is possible.

4.2. FRAMES OF REFERENCE

Through the classification of frames of reference a second design guideline can become imaginable. In this respect, more detailed investigations are needed to analyze not only how virtual presentation schemes are seen under different frames of reference, but also where the references are placed within the scenery and how they relate. Thus, relationships between AR objects, their mounting and their meaning, are to be expressed.
General types of mounting are:

- Location-fixed presentation: Objects are embedded at a fixed position in the environment, for example, navigational arrows.
- Body/vehicle-mounted presentation: The virtual representation remains in a position relative to the user, for example, the driving dynamics bar presentation or the 3D warning direction arrow.
- Head-mounted presentation: Information that always has to stay in the field of view of the driver is displayed in that way.
- Glance-mounted presentation: Preferably high priority information to quickly gather a user's attention is displayed through this kind of mounting.

Information presentation is constrained by its relation to the environment. Several studies have been conducted to determine which kind of information is to be displayed intuitively, for what kind of content, and in which frame of reference to the user.

Thomas *et al.* (1999) investigated how frames of reference (FOR) affect situational awareness. Situational awareness concerning mobility can be split into a navigation task requiring global awareness and a driving task requiring local guidance (Barfield *et al.*, 1995). Global awareness is the knowledge about the route to the destination. Local guidance includes tasks that involve controlling the vehicle and knowledge about the environmental situation. Local guidance focuses on understanding the spatial relationship between a controlled object and its immediate surroundings. Wang (2004) has also compared egocentric and exocentric navigation assistance as a function of viewpoint tethering. He states that global awareness of the environment improves with the length of the tether whereas local guidance performance deteriorates. Milgram and Kishino (1994) give the taxonomy of mixed reality presentation schemes ranging from egocentric to exocentric, suggesting the use of egocentric visualisations for local guidance. Experimental results of Barfield *et al.* (1995) have also consistently shown that local guidance is supported best by egocentric visual information.

Yeh *et al.* (1998) state that travel benefits are achieved with an egocentric FOR which provides a natural compatibility between perception and control such that the display viewpoint is identical to the axis of control. In their survey, Yeh *et al.* state that tasks involving the understanding of objects' locations within the environmental space benefit from the use of the exocentric viewpoint. This result becomes clear upon viewing the immersed display in an egocentric reference frame, which requires mentally piecing together various 'snapshots' of the environment taken from different perspectives of virtual space to form a 'big picture.' Conversely, the exocentric display presents one global view from one 'permanent' angle.

Mapping these experiences to mounting points for AR-based presentation schemes and thus to design guidelines for integrated automotive concepts, will not immediately determine a perfect solution, but several hypotheses can be postulated for further research.

Several evaluations with respect to FOR resulted in the finding that global situation awareness is improved through more exocentric views on the scenery. It was found that objects and places to be indicated in larger distances should use exocentric presentation schemes. Those schemes are best placed in a screen-fixed presentation as they do not have a direct mapping to the real outside environment. When objects such as obstacles or crossings come nearer to the driver, the view should change to an egocentric point of view. Tethered view-point functions appear promising for this transfer of presentation. Navigational arrows which are directly migrated from the real world to the virtual world should then use the symbolic bird's eye presentation scheme in a near focal distance for large distances, and should become location-fixed (conformal) as the point approaches.

To support local guidance, conformal egocentric presentation schemes should be used. Thus, our approach for stabilisation tasks, the braking bar scheme, still fits into general recommended presentation guidelines, because it relates to the near environment. Its mounting relative to the car immerses the coupling as an aid for stabilisation. This presentation scheme will not interfere with any other scheme in an integrated system except new issues for AR-based driver assistance.

The suitability of warning schemes which use arrows needs to be investigated before adoption. The car-mounted egocentric appearance is suited to its application area of enhanced situational awareness. The mounting located near the user at the front bumper contributes to car related information. An understanding of the relationship between guidance of attention and navigation tasks does not yet exist. Can these be clarified intuitively, or how can more implicit knowledge about an object's semantic meaning be aggregated? Any navigational location-fixed arrows are flattened shapes lying on the streets and thus relate to the world's two-dimensional, sometimes bent (hills, et cetera) surface. Our 3D arrow uses a 3D shape and can point in any direction, even upwards (that is, if a bridge is too narrow). The differences in appearance can convey implicit content about the meaning of the scheme. Also, the coupling with spatial sound increases the signalling of danger, and makes this scheme appear more as a warning scheme than a navigational aid. Detailed results will require studies on a combined system incorporating both presentation schemes concurrently.

It is useful to define guidelines for such a combined approach before conducting user studies. Only through this approach can elementary mistakes in user interface design be avoided. Frames of reference and mounting points can define general guidelines for the design of such interfaces. Each interface in the whole system must transfer its implicit meaning not only through shape and behaviour but also through its mounting and relation to the environment. Arrows in particular, which often provide outperforming solutions for single tasks, become problematic when applied in integrated systems. Design processes of arrow schemes must take into account the extent to which they relate to the environment. Arrows related to objects or places in the far distance should use shapes of a more symbolic character, while presentation schemes related to nearby objects can use the whole width of an egocentric frame of reference. Visual and transformational mounting of these presentation schemes has to support the corresponding frame of reference and cannot be used to signify further implicit information.

5. Summary

Arrow-based AR presentation schemes provide a good metaphor for various applications. In automotive environments we have investigated several presentation schemes, sometimes also using arrows to support drivers in their task of driving. AR based information presentation does reduce the amount of time that would otherwise be needed to gather information from secondary displays or other sources. Compared to in-car displays, AR presentation schemes significantly reduce or completely dispense focal adaptation time. On the other hand, AR schemes deepen questions and problems pertaining to information overload, perceptual tunnelling and cognitive capture. Therefore, a driver's workplace was investigated before introducing some of our contributions to safety into the automotive sector.

Our current set of solutions is part of a developing process aimed at developing a fully integrated human-centred man-machine interface. Therefore, we have subsequently investigated issues towards the development of such systems. Issues pertaining to the design of AR presentation schemes have been our chief focus. Here, the frequent use of arrows appears to become a problem in itself. Discussion of our concepts in respect to tasks of mobility and frames of reference are promising steps toward an integrated approach.

Our further research will investigate the hypotheses of AR design in respect to the frames of reference stated here. Future systems will also incorporate priority queuing concepts to keep information density under a certain threshold.

Augmented Reality in the automotive domain is a rewarding area of application as modern cars, tracking and presentation technology begins to enable new human machine interfaces. Prior to such interfaces reaching the market, issues ranging from legal regulations to successfully modelling a single human's spatial understanding need to be addressed.

5 MIXED REALITY IN EDUCATION/LEARNING

VISUALISING FUTURE CITIES IN THE ETH VALUE LAB

New Methods for Education and Learning

REMO BURKHARD AND GERHARD SCHMITT
ETH Zurich, Switzerland

Abstract. This article discusses how the use of complementary visualisation techniques can contribute to improve planning, understanding, and communication of future cities, especially when different stakeholders are involved. First, it describes a framework to structure visual representations. Second, it introduces the 'ETH Value Lab' as a tool for designing future cities. Third, it introduces two applications that can be used for two urban planning processes: planning and project management and visualising neighbourhoods. Finally, it shows scenarios for education and learning. This article is relevant for urban planners and visualisation researchers, because it points to the emerging field of visualising future cities and for professors, teachers, but also school administration and ICT-experts who want to invest and use state-of the art mixed reality infrastructures for teaching and research.

Keywords. Visualising Future Cities, Value Lab, Mixed Realities, Knowledge Visualisation, City Engine.

1. Introduction: Visualising Future Cities

Visualising knowledge so that it can be better understood, discussed, or communicated is a long-standing objective in different fields including in urban planning and urban design. Today, urban planning and urban design still rely mainly on static representations, such as maps, conceptual drawings, key visuals, and physical models. Most of these static visualisations are currently created with computer-based tools (for example, CAAD). However, in spite of progress in defining standards such as STEP (Standard for The Exchange of Product model) and IFCs (Industry Foundation Classes – class library), the computer is still often only used as a drawing board. Relatively few students and practitioners use the computer to visualise dynamic data or scenario

X. Wang and M.A. Schnabel (eds.), Mixed Reality in Architecture, Design and Construction, 205–218.

simulations. One example: Semi-automatically generated 3D city models exist, yet are not used in planning in an integrated manner, even if they are helpful for analysing scenarios, designing volumetric studies, managing complexity, and visualising scenarios for urban retrofits.

1.1. WHY VISUALISING FUTURE CITIES?

First, due to their complexity, we have difficulties in perceiving, representing, and communicating future cities. Future Cities, especially mega cities have to be understood as a dynamic system – a network that bridges different scales, such as local, regional, and global. This network comprises several dimensions, for example social, cultural, and economic. Due to this multidimensional character of a future city and its network properties, we have difficulties perceiving it. Both researchers and the public cannot answer simple questions such as: Where does a neighbourhood, a city or a mega city start and end?

Second, we do not have enough experience to manage participatory planning processes, for example to establish a mutual vision, or to map the desires of the involved participants.

Third, we have not yet mastered the challenge of visualising non-physical contents. How can we map functional relationships and interdependencies of urban ensembles in mega city regions? Which methods allow us to visualise long-term planning processes? How do we map clusters of knowledge in mega city regions?

1.2. WHY FUTURE CITIES?

Cities like Shanghai, Beijing or Sao Paulo, just to mention a few, are rapidly growing. Prognoses state that 90% of global population growth will occur in cities between now and 2030. Infrastructures and the environment have to be adapted to the changing demands and new urban development strategies have to be elaborated. In 2007, for the first time, more people lived in cities than in the countryside, and until 2015 the number of cities with a population in excess of 10 million people will grow from 300 up to 560 so that 350 million people will live in mega-cities (Burkhard *et al.*, 2007). One main reason for this above average growth lies in the economic attractiveness of metropolitan regions. If we take our environmental responsibility seriously we have to make these future cities and the redeveloped existing cities more sustainable. Resulting challenges are: How can we optimise infrastructures (for example, transportation, water, communication systems) and buildings through new concepts, new technologies and new social behaviours to cut down CO_2 emissions, energy consumption, traffic load, and to increase the quality of life?

To answer these questions dynamic and interactive visual representations support various purposes such as the analysis, design, planning, management, surveillance, or maintenance. It is important to understand that 'visualising future cities' means more than the use of Geographic Information Systems (GIS), Computer Aided Architectural Design (CAAD), or reality-based modelling. Just as important – or even more important – are emotional visualisations (for example, stories, key visuals, movies) that help to communicate, establish a shared vision and a collective desire. Finally, new approaches are needed to implement future cities with heterogeneous stakeholders in cooperative city planning processes. Thus, it is a truly interdisciplinary challenge, where many different skills need to be combined: Computer Science, Environmental Studies, Sociology, Design, Communication, Urban Planning, Strategic Management, Architecture, and Aesthetics, plus feedback from the general public. And this challenge will have significant impact on education and learning which will be discussed in this paper.

2. Theory: The Use of Complementary Visualisations

Most of our brain's activity deals with processing and analysing visual images. To understand perception, we have to know that our brain does not differ greatly from our ancestors, the troglodytes and at that time, perception helped for basic functions, for example for hunting (motion detection), seeking food (colour detection), or applying tools (object shape perception). Since then, visual representations have served a variety of functions such as addressing emotions, illustrating relations, discovering trends and patterns, getting and keeping the attention of recipients, supporting remembrance and recall, presenting both an overview and details, facilitating learning, coordinating individuals, establishing a mutual story, or energising and motivating people.

In this section, the main visualisation types are distinguished. They are derived from the practice of experts in visualising future cities, such as urban planners, cartographers, architects, marketing experts, writers, and film makers. The visualisation types can be structured into seven groups: Sketches, Diagrams, Images, Maps, Objects, Interactive Visualisations, and Stories.

Sketches are atmospheric and help to quickly visualise a concept. They present key features, support reasoning and arguing, and allow room for individual interpretation. Sketches are heavily used by architects and urban planners for analytical and design tasks and to communicate ideas or visions. Specific types are instant napkin sketches, for example to explain the way to a specific place, which is used in Tokyo, due to the fact that they do not have street names and numbers.

Diagrams are abstract, schematic representations used to explore structural relationships among different parts by denoting functional relationship. Diagrams explain causal relationships, reduce the complexity to key issues, structure and display relationships. Quantitative diagrams are used to visualise statistical information or economic indicators, such as climate curves, population, and growth et cetera. Diagrams are used by architects and urban planners to visualise elements of cities, such as functional zones or flows of persons. Diagrams are also used to visualise the different phases of a planning project.

Images[1] are representations that can visualise impressions, expressions or realism. An image can be a photograph, a computer rendering, a painting, or other format. Images catch the attention, inspire, address emotions, improve recall, and initiate discussions. Images are instant and rapid, instructive, and they facilitate learning. In the context of mega city regions we can benefit from satellite images of mega cities. Then, in the marketing of a mega city region, photographs are often used to visualise highlights of a city, for example, monuments, restaurants, shopping streets, business centres, museum districts, and events. Additionally, images are sometimes used to explain more abstract concepts, such as quality of living and benefits for companies. Then, mega city regions can be communicated by means of visual metaphors. Such visual metaphors and analogies support recall, lead to 'a-ha' effects, and to discussions. An example is the strong image of an elderly man in a traditional costume in a beer garden working with a high-tech laptop. This image illustrates that Munich is both a site with traditional roots and a high-tech industry.

Maps represent individual elements – for example, roads – and in a global context – such as a city. Maps illustrate an overview and details, relationships among items; they structure information through spatial alignment and allow zoom-ins and easy access to information. Maps generally have a scale that determines the size of an object represented on the map in relation to its actual size. Some maps are not scaled, for example the tube map that uses a visual system that distorts the real distances to obtain a more readable map. The features on a map depend on the map's purpose: a road map displays roads, a tube map shows the tube system, and thematic maps represent thematic entities such as the thematic similarities in the top right window in Figure 1. Further examples of maps are interactive satellite maps (Global Positioning System, GPS) – combined with superimposed layers of location based information (for example, restaurants, shops, history of a building) – for car drivers or users of mobile devices.

Objects exploit the third dimension and are haptic. They help attract recipients (for example, a physical dinosaur in a science museum), support learning through constant presence, and allow the integration of digital interfaces. Many cities have a wooden three-dimensional model of their city, but

---

[1] In the English language we have to distinguish between 'picture' and 'image.' Pictures are more physically manifested representations, whereas images have stronger connotations with mental images. In this article, we only use the term image and refer to both, physical and mental images.

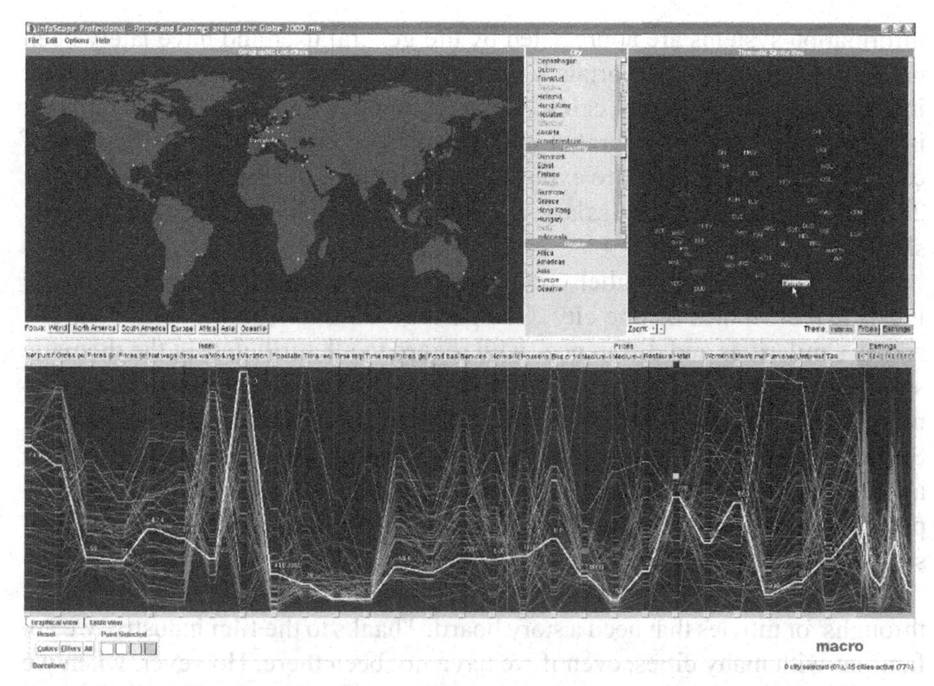

Figure 1. An interactive visualisation allows visually analysis a high-dimensional dataset. Above, Barcelona and Hong Kong are selected and the slider 'Hotel' is filtering the dataset (Copyright Macrofocus GmbH, www.macrofocus.com).

often only of their core cities. An impressive model is the model of Shanghai. The creation of such models is expensive and time consuming. Here the new rapid prototyping tools help. Despite all the wonderful possibilities of virtual reality applications, the qualities of a physical model still attract us more and are more suitable for gaining an overview and understanding spatial relationships. While such models are helpful and attractive, they also face a problem when it comes to mega city regions. Here the very large territory leads to a problematic relationship between the heights and lengths – models of mega city regions then more closely resemble a landscape model than a city model. As soon as we need to filter or work with the model and switch on or off different layers of information, a virtual model can become more powerful. Such a virtual three-dimensional model annotated with semantic information (for example, Google Earth) can be used to simulate all kinds of information (for example, weather, history, people, cars, et cetera) or to simulate temporal data, such as potential development scenarios.

Interactive visualisations are computer-based visualisations that allow users to access, control, combine, and manipulate different types of information or media. Interactive visualisations help catch the attention of people, enable interactive collaboration across time and space, and make it possible to represent and explore complex data, or to create new insights. Interactive geographic information systems are appreciated by the general user and have lately been used by companies or portals as an orientation layer to map additional information, such as hotels, cafés, or instant mapping of the amount of rain that is falling, and can be integrated into websites or into onboard computers of cars. Figure 1 shows a tool to explore and compare cities according to roughly 20 criteria, such as average salary, average price for a bus ticket, or population size.

Each line in this 'parallel coordinates' view connects the values of the individual attributes of one city. Comparing two lines (and thus two cities), shows similarities and differences with regard to all attributes. In the thematic similarity view each city is visualised with a dot. The closer two dots are, the more similar the two cities are with regard to the criteria, and vice versa. Sliders can be used to interactively filter the dataset. This kind of visualisation allows exploring a larger amount of structured data. It is good for analytical purposes, but too complex for communication to the general public or different stakeholders.

Another subgroup of this type includes interactive animations: 'fly-throughs' or movies that need a story board. Thanks to the film industry, we are familiar with many cities, even if we have not been there. However, when we watch a movie of the city we live in, we often realise that it is a distorted reality. One reason why many guests from India visit Switzerland is that the 'honeymoon in Switzerland' is a key element in the Indian film industry. That is the reason that a great number of Indian tourists come to visit the original sites. Similar sectors that will become more important in the future are virtual cities and the gaming industry. Another example of interactive visualisation is augmented reality, which means superimposing relevant information in real time on windscreens of cars or special eye glasses.

Stories and mental images are imaginary non-physical visualisations. When we think of a city we automatically remember stories associated with that city. Stories are a very effective method to share different impressions and experiences. One format that captures personal experiences is a diary. Weblogs are a new form of public diary. They exist for cities and probably soon for mega city regions as well. Mental images are another type. Kevin Lynch (1960) pointed to the spatial perception of city users and associated mental maps, which consist of five elements: (1) paths on which people travel, such as streets, trails; (2) edges and perceived boundaries, such as walls, buildings, and shorelines; (3) districts, such as sections of the city with a specific identity or character; (4) nodes, such as focal points, intersections or loci; (5) landmarks, readily identifiable objects, which serve as reference points. Architects often envision and discuss scenarios. In order to explicate these ideas, architects might rediscover the power of narrative texts to explain their concepts to the general public. Here we can learn from novelists. One example is Italo

Calvino's novel (1974), in which the Mongolian emperor Kublai Khan was too busy governing his empire to travel, so he asked the explorer Marco Polo to describe the cities he had seen. The short descriptions of the 55 cities combine facts and tales Marco Polo had heard about the city regions and cities. Some of them only existed in the imagination, such as an underground city of the dead.

Storytelling is of course also used to explain how to get from one place to another. In some situations such narrative descriptions can be more suitable than a map. For example, they help to navigate in the narrow and maze-like streets of a North African Medina. The key to orientation here is to remember the different souks (market places).

This section introduced seven visualisation types. The next section will present the ETH Value Lab: A tool that allows work with complementary visualisations.

3. Tool: ETH Value Lab

The ETH Value Lab is a new teaching and research facility for collaborative knowledge creation. One key application area where the Value Lab will be used is the design of future cities.

The ETH Value Lab (Figure 2) is located in the Information Science Lab building within the Science City ETH and offers new ways to work with digital data.

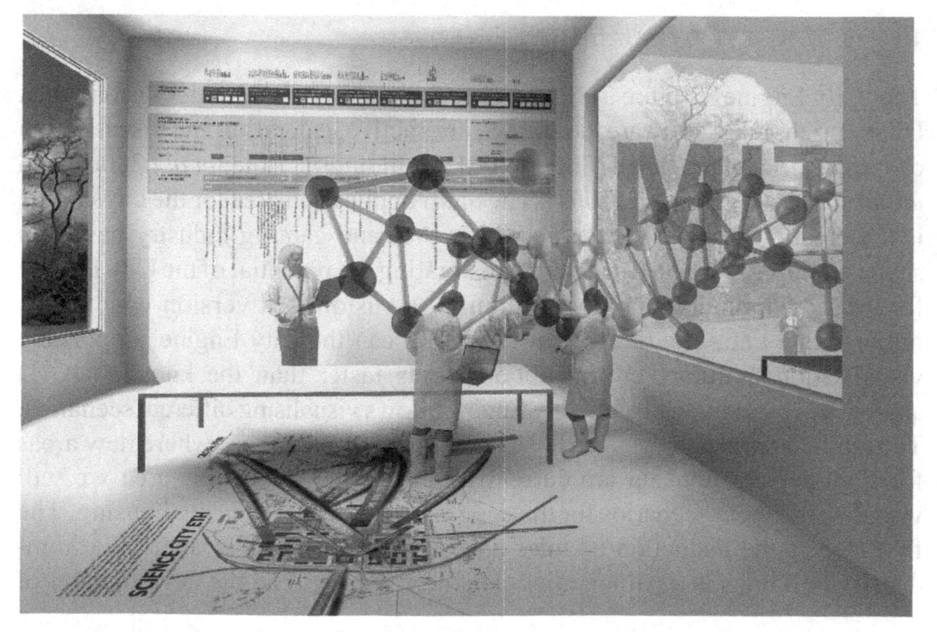

Figure 2. Value Lab conceptual image of mixed reality (Copyright vasp datatecture GmbH, www.vasp.ch).

The ETH Value Lab (approximately 6 m wide, 10 m long, 6 m high) consists of two interactive tables (each one consisting of a 65 inch LCD screen) and 4 LCD-screens (each one 82 inch) forming a two-by-two matrix fixed to the wall. The system is tightly coupled and connected via fibreglass to a central server in the basement of the building. Additionally, two HD projectors and a Dolby surround system are installed. We have chosen a setting with LCD-screens and not a rear projection system because of the disadvantages like daylight sensibility, noise, service costs for lamps, amount of consumed space and also because of the limited size of the room.

The planned completion of the ETH Value Lab is in mid 2008. It will be integrated into research and teaching within the ETH and with partner universities, but also with partners from industry. The setting is customised for interactive sessions with up to 15 users. It will be used for the simulation and cooperative planning of future cities. However, the setting is not restricted to this area. The ETH Value Lab also allows analysis of large data sets, monitoring of transportation networks, real-time visualisations, distributed real-time rendering, interactive screen-design and many other applications.

4. Value Lab Applications for Visualising Future Cities

In this section we introduce two tools that are currently being developed for the design of future cities in the Value Lab.

4.1. FUTURE CITY VISUALISER

The 'City Engine' (Figure 3) has been developed in the computer vision lab at ETH Zurich. It is a software tool for the interactive generation and visualisation of virtual cities (Parish and Müller, 2001), buildings (Müller *et al.*, 2006), and urban spaces (Ulmer *et al.*, 2007). The main application area for the city engine is the creation of virtual cities for the movie and gaming industry based on shape grammars. We are currently investigating the potential of the City Engine for urban planning and are programming a customised version which runs in the Value Lab. The research question is: 'can the City Engine support the visualisation of future city neighbourhoods faster than the known CAAD approaches?' Additionally, the City Engine allows visualising different scenarios. To test the approach we have chosen a real site in Singapore where new areas for high-density housing are currently being planned. In this area we will visualise different scenarios for the development of high-density housing. The first step is the deduction of rules and shape grammars by analysing high-density buildings in existing cities. The next step will be the development of new high-density building designs and the visualisation of entire neighbourhoods. This will enable us to compare the different scenarios with leading experts such as urban planners, authorities, architects, and local inhabitants.

This first application is a mixed reality approach which will lead to new formats in education and learning.

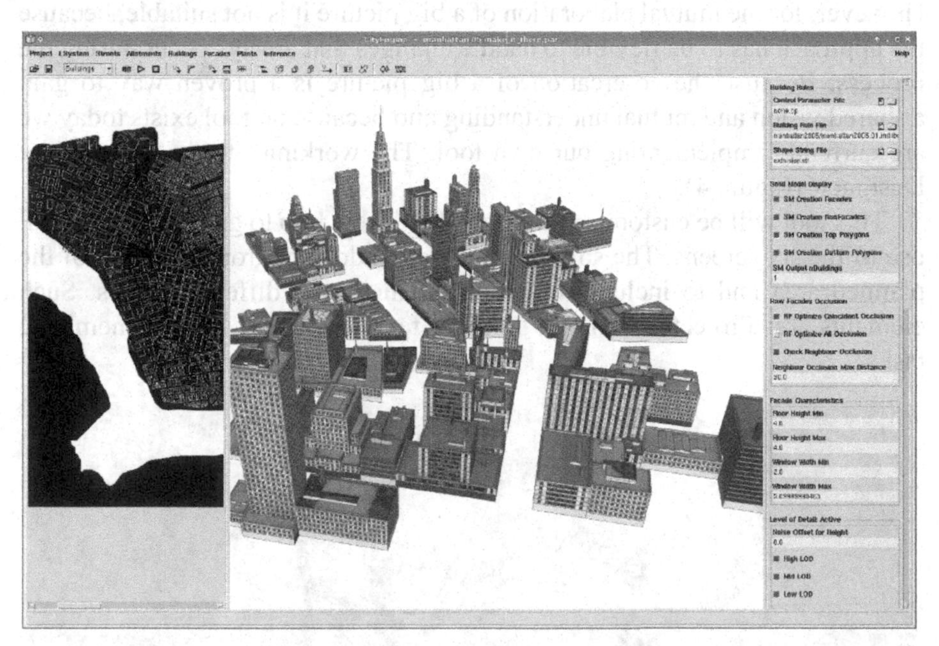

Figure 3. City Engine: software for simulating future city districts based on shape grammars (Ulmer *et al.*, 2007).

4.2. FUTURE CITY DESIGNER

In cooperative planning it is very challenging to establish a shared understanding among the involved partners. In this case, the coordinative functions of visual representations are a significant help. A static strategy map or roadmap for example can facilitate aligning all partners to one mutual 'big picture'. While a static map designed by one author already offers support, the most valuable effects can be measured when the map is designed by a group in a moderated process. If a group of people develops a visual representation in collaboration, they will accept it and adhere to it more strongly. In contrast, if they are not actively involved, they will read it and most likely forget it, but certainly not enthusiastically share it with peers. We have experimented for more than 5 years of using current software applications for the interactive creation of such co-created maps in workshop settings of around 15 people. On the one hand, Microsoft PowerPoint or similar tools do not allow the necessary graphic flexibility that is needed to design an appealing visualisation; on the other hand, Adobe Illustrator or similar tools are not suitable when it comes to the projection with video projectors. Constant zoom-ins and zoom-outs kill the flow of the workshop. Finally, using non-digital tools, for example, flip charts, is still a proven method for brainstorming sessions and the collection of ideas. However, for the mutual elaboration of a big picture it is not suitable, because the approach has to be flexible to handle changes that always occur during the process. Because the co-creation of a big picture is a proven way to gain a shared vision and mutual understanding and because no tool exists today we are currently implementing our own tool. The working title is 'Future City Designer' (Figure 4).

This tool will be customised for the ETH Value Lab to guarantee maximal use of the six screens. The software allows to add background images of the planned area and to include individual elements on different layers. Such elements are: Projects, reference images, texts, sketches, project members, tasks.

Figure 4. Future City Designer (Copyright ETH Zurich, Chair for Information Architecture, www.ia.arch.ethz.ch).

A sample workflow of a workshop is as follows: The group starts the workshop by choosing the context (for example, Singapore) and agreeing on the planning areas for new projects (for example, five areas in Singapore) by defining the respective zones. To do so, they mark the area with their fingers on the interactive table. Then, they brainstorm on projects that could be developed in the five areas. To do so, they use the sketching tool to mark zones or add project icons that allow adding metadata to the projects. A project can be a new landmark or an event. Then, they load reference images to visualise the idea (for example, three photos of similar landmarks or similar events) to achieve a shared understanding in the group. The group discusses the feasibility and relevance of the project and might enlarge the project icons of the projects that are more important or they delete projects. They zoom into the project and add tasks that need be done to implement the project. Finally, they use the people-allocation functionality to add people to tasks or link tasks to people. We expect that the effects of using this application are a shared and better understanding, more sustainable planning, higher motivation through personal engagement, better decisions, and realistic work-break-down structures.

5. Mixed Reality in Education and Learning

The ETH Value Lab represents the latest generation of the ETH large-scale simulation environments that began in 1998 with the establishment of the VisDome (Figure 5), a 10-m diameter immersive visualisation space with stereo projection and interactive modelling and visualisation software under the dome of ETH Zurich's main building. It placed a most advanced Virtual Reality environment in a unique architectural space, designed by Sumi and Burkhalter. It is still in use and was the prototype for several similar spaces worldwide. It allows for the interactive design and experience of spaces in architecture, engineering, medicine, and the natural sciences. It turned out to be most effective for presentations and individual work, less for collaborative design.

Figure 5. VisDome – a typical example for individual design and collaborative experience of results. Simulated view from the control space. Interactive presentations occur in the centre, three projectors float under the lighting ring (Copyright ETH Zurich, www.ethz.ch).

The follow-up to the ETH VisDome was the BlueC (Figure 6), a next-generation immersive cave. Rather than being restricted to the typical cave limitations, BlueC introduced a fundamental innovation: users not only viewed and manipulated objects in 3D, but 16 cameras placed outside the cave recorded the users inside the cave, calculated the 3D model in real time, added textures, and compressed the model, thus creating a real life avatar that could be sent to re-emerge in other locations. The technology behind the BlueC is extensive, in that the surrounding panels consist of a special industrial layered glass that could be switched between being transparent and opaque up to 60 times a second. This allowed for projection from the outside during the opaque phase and the recoding of the users during the transparent phase. The high frequency made the switching invisible to the human eye, so that projection and recording seem to appear concurrent. The synchronisation of the cameras, the projectors and the glass behaviour posed a technological challenge, the close to real time 3D scanning, model building, and compression led to major software advances. Overall, 14 PhD students from several disciplines created the environment.

Figure 6. Fisheye view of BlueC. Projectors and cameras are placed outside the glass panels, LED lighting devices on the floor (Copyright ETH Zurich, blue-c.ethz.ch).

Applications of the BlueC include financial modelling, architectural and mechanical design. Andrew Vande Moere (2004) used BlueC to develop his Infoticle concept. Typical interactions occurred between several users or between users and virtual objects in the environment.

The BlueC soon was developed further into a more architectural space. Kai Strehlke designed the so-called 'Red Hell,' named after the colour surrounding the installation. A space holding up to 30 people with a floating back projection wall at one end held was used for design critiques with the advantage that physical models could be present in front of the projection wall. The recording cameras are placed in the red walls on two sides and in the ceiling. The Red Hell can project the 3D and textured avatars recorded in the BlueC and vice versa. As the avatars can be displayed anywhere, BlueC and Red Hell could also act as 3D mirrors. This led to new applications, such as fashion modelling and virtual dressing.

Figure 7. Fisheye view of the Red Hell. Projection occurs from the back of the screen, the red walls and the ceiling hold the recording cameras (Copyright ETH Zurich, blue-c.ethz.ch).

By 2003, collaborative design already had a 10-year history at ETH, and BlueC and Red Hell offered new technical opportunities. However, the underlying technology and software had become very complex and needed constant maintenance by specialists. This placed a barrier – between designers and technology – to the seamless integration of the otherwise perfect design environment.

At the same time, ETH experimented with other advanced interactive projection devices. Students in the Department of Pharmacy and Applied Biosciences, used wall mounted displays and large displays built into tables. Their mixed reality environment consisted of wet laboratories, the physical library, and the interactive modelling space. In addition, they participated in telepresence lectures with their colleagues at the University of Basel.

For the future, we foresee the mixed reality environment for regular design studios as an adaptation of the environment that has emerged at ETH over the last decade: a large central space in which paper models, city models, artistic performances, discussions and presentations take place, surrounded by smaller spaces in which groups of 15–20 students interact with the assistants and professors and use their individual laptops and modelling devices. For special studios and interactive simulations of results, the Value Lab will be the dominating environment as it offers software and hardware that will become affordable and mainstream only in the coming decade. Definitively, the technology and the changed teaching and learning environment will evolve the pedagogy and learning progress. Although it may be difficult for many students to apply and program the new equipment, it is necessary that students undertake this task, as they will be the next generation of users and teachers.

6. Summary

This article discussed how the use of complementary visualisation techniques can contribute to improve planning, understanding, and communication of future cities, especially when different stakeholders are involved. First, it described a framework to structure visual representations. Second, it introduced the 'ETH Value Lab' as a tool for designing future cities and two applications: the City Engine for the visualisation of future city neighbourhoods and the Future City Designer for cooperative planning processes. Then it proposed scenarios for learning and education. The article aimed at introducing the emerging area 'Visualising future cities' as a key area for teaching and research and the 'ETH Value Lab' as a new tool for innovative and fruitful research and teaching.

INTERPLAY OF DOMAINS

New Dimensions of Design Learning in Mixed Realities

MARC AUREL SCHNABEL
University of Sydney, Australia

Abstract. There is a distance between the idea of a design in the imagination and its representation, communication and realisation. Architects use a variety of tools to bridge this gap. Each tool places different demands on the designer and each, through inherent characteristics and affordances, introduces reinterpretations of the design idea, thus imposing a divergence between the idea and the expression of the idea. Design is an activity that is greatly complex, and influenced by numerous factors. Most researchers of Mixed Realities (MRs) have focused on their use as presentation or simulation environments. It has been suggested that MR can empower designers to express, explore and convey their imagination more easily. For these reasons the very different nature of MR with its unique properties may allow architects and learners to create designs that other instruments do not offer. There has been inadequate exploration in the use of these realms for the acts of designing, as well as in educational contexts of design-learning.

Keywords. Learning, Communication, Interplay, Design Generation, Design Exploration.

1. Introduction

Architectural design within Augmented, Mixed and Virtual Realities has been widely used as a method of design simulation and presentation. Educational and professional settings employ these realms successfully to study, communicate and present architectural designs. The rapid development of digital tools over the past decades has had profound impact on architectural education and the ways in which architects create, converse or appreciate three-dimensional spatial environments (Koutamanis, 2000). Numerous publications illustrate the impact that digital media have had on design studios and propose solutions

X. Wang and M.A. Schnabel (eds.), Mixed Reality in Architecture, Design and Construction, 219–226.

for multi-media design studios and ways in which to make use of Mixed Realities (Maver, 2002). Dave (1995) investigated distributed design studios, Wenz and Hirschberg (1997) studied collaborative design within remote collaboration, while Hirschberg *et al.* (1999) analysed patterns of communication within digital design studios. Mixed Reality (MR) often became an instrument to assess design alternatives and final design solutions (Achten, 2001). Yet, none of the authors looked into the comprehension and conception of designing within MR.

A particular form of design studio emerged in the early 1990s that investigated various possibilities that digital media and Virtual Environments (VE) could offer to the learning and exploration of architectural design. These 'Virtual Design Studios' (VDS) defined *virtuality* as acting while physically distant or as acting by employing digital tools (Maher *et al.*, 2000; Schnabel, 2002). It became apparent that the next logical steps to develop these design studios were to combine real and virtual environments in an MR experience. Mitchell (1995) also refers to the need for an ongoing evolution of digital design studios towards a fully integrated studio where the borderlines between realms are dismantled. In the same way, Chen *et al.* (1998) suggest that human-human interactions could take place within and throughout conventional and computer systems of a new type of virtual studios, instead of through or external to them, as it did in some of these digitally supported studios.

2. Virtual Dimensions

MRs have to be studied together with VEs to comprehend the influence that virtual aspects have on a realm where real and virtual elements merge into a new dimension. Similarly to MR, VEs were originally embraced by archi tects for design concept presentations. As computing advances, increasingly sophisticated interaction and design possibilities are needed and supported (Hendrickson and Rehak, 1993). According to Maze (2002) however, VEs are seldom used for creation, development, form-finding and collaboration of architectural design. Likewise Immersive-VE (IVE), which enables active and real-time interactions with design, has not yet been used widely in the design process. Schnabel and Kvan (2003) report that IVE offers new opportunities and solutions to architectural design problems through involvement in a three-dimensional (3D) medium. They argue that, via employment of IVE to create and realise ideas, the architect is challenged to deal with perceptions of solid and void, and navigation and function, without translations to and from two-dimensional (2D) media. Furthermore, they suggest that VE empowers designers to express, explore and convey their imagination with greater ease. For these reasons, the very different nature of IVE allows architects to create designs that reflect the three-dimensionality of architectural design to a greater precision than in 2D realms. Virtual Reality (VR) is a constructive tool that supports the design and communication process (Davidson and Campbell, 1996). Compared to conventional computer-aided design (CAD), designing within IVE does not present with the typical lack of collaboration and communication as noted by Kvan *et al.* (2000). The exploration of space, volume and location is enhanced and site-specific problems are not only better recognised, but possibilities are also better investigated, both of which a normal design process cannot offer (Campbell and Wells, 1994). Users of IVE can change their viewpoints and escape gravity, all the time remaining 'inside' the model without having to translate scales or dimensionalities. The research found that designers prefer to work three-dimensionally because every creation within IVE is a place experienced directly through movement and interaction parallel to real world familiarity.

The research findings of a design studio held in an IVE show that this realm produces different architectural expressions and exploration of form and gestalt from those explored with 2D tools (Schnabel, 2004). The design proposals illustrated that the 3D space is explored and used extensively in order to create innovative schemes. This proves that designers can successfully use the medium to create and communicate architectural structures within a normal studio setting. Thus, the process of collaboration and design is enhanced and communication between designers is more focused on the subject itself.

In his research, Schnabel demonstrated that employing VE as the medium for the design process enhances the designer's perception and understanding of 3D form, volume and space. This is true not only for purposes of presentation or simulation, but also for different stages of the design process itself. From the results, it was also identified that a direct translation of information from VE into other real media is potentially problematic. However, similar to the conclusions of Yip (2001), it was found that re-representation and translation into other realms contributed to the quality of the overall design process.

Schnabel and Kvan (2003) also point out that despite the advantages of VEs, a re-representation within other media – or a mixing of realms – will lead to a deeper understanding of spatial design. Hence, an MR will contain intersections of properties of the real and virtual realms, and allow designers to interact within this MR to create, explore and communicate their designs.

An MR offers designers an instrument that allows conceptualisation of design ideas in a finer way, whereby digital 3D models are generated with immediacy similar to physical reality, constructed to improve the perception of designs developed by drawings. Through its involving qualities, MR provides immediate feedback to its users, which is not possible within CAD or traditional design media. Subsequently, architects can negotiate their own design despite the technology used and the abstractness of MR. This process of design promotes the convergence of the idea and the design intention in a manner that is closer to a normal interaction with 3D media. In that sense it relates to a 'human' interaction.

3. Mixed Domains

In their study, Underkoffler and Ishii (1999) report that MR is a practical medium to aid the design and communication process, and to establish collocated presence for a joint experience in design reviews. This leads to the significance of a shared learning experience, which is an essential part of the curriculum in architectural design. Students need to learn the common vocabulary of designing, its syntax and grammar of communication. Current digital design tools do not make sharing particularly easy. Perceived usability does not originate from a design-related background; rather, it has been adapted from other domains such as engineering. Interfaces in desktop design tools were originally designed to accommodate a single user only and have recently been extended to facilitate the need for 'shareability.' This becomes crucial in the context of architectural design, where peripheral data are used extensively and described through spatial abstraction. Through this, the needs and aspirations of a variety of stakeholders are negotiated and represented. This provides an opportunity to gather data on the necessity of design communication and collaboration between the participants through the means of an interactive medium like MR.

As Schnabel and Kvan (2003) have reported that designers within a VE gain an enhanced understanding of spatial relationships, the designers' *sense of presence* has an important role to play, allowing a finer interaction with the design. Since MR permits a blending of virtual entities with a real environment, it can as a result increase the comprehension of the design proposal. MR creates the ability to move around in space, merging real with virtual models and designs, as well as to change scale. Thus the understanding of the design and its relationship within the architectural context is enhanced, site-specific contexts are better recognised and a variety of options can easily be investigated.

Imagination is part of the process but it is limited by some constraints of communication media and representation. Designers therefore need to apply a variety of tools to overcome these limitations and to allow them to express and communicate their ideas. Schnabel (2002) found that designing within all dimensions of space leads to a richer exploration of the design. Within the different properties of MR designers are also empowered to express, explore and convey their design three dimensionally thereby consequently reducing the divergence between idea and creation. That will ultimately lead to an improvement of the overall design process and its communication. Due to the novelty of the instruments and realm however, there might be some factors that influence the design outcome that are caused by the use of the new medium. With time and ease of use this will be overcome without doubt and will not influence the act of designing.

4. Innovative Learning

MR as a teaching tool for architectural design offers fascinating new possibilities. Students and teachers can explore a variety of theoretical and practical frameworks in order to understand dynamically and spatially complex relationships. Earlier limitations within either physical or virtual realms are reduced and advantages of both can be merged into one environment. An architectural design studio that employs MR as a tool, allows novices and experts to communicate and collaborate instantaneously. Hence the participants explore, communicate and understand spatial issues in a new way. Designers are able to work interactively as every object within the simulated environment is experienced through movement, interaction and immediate feedback. These possibilities offer a different 'conversation' with the design that otherwise is not obvious or possible. Spatial and architectural issues can be addressed in a manner akin to the real world in which MR enhances the translation of the designers' intention. A certain credit has to go to the experiences that were encountered by the use of the technology and the abstractness of any given model.

Today's common computer hardware and software enable architectural design studios to employ an MR-system and its components easily. There are no longer major technical overheads that have to be dealt with. This enables collaboration between remote partners and a translation of theoretical and practical architectural design issues to remote locations.

Following the arguments proposed by Bosselmann (1998) and (Hack and Canto, 1984), it is important for architects, in the early design stages to use a medium that reflects the complexity and interactivity of the site and the proposed design. Using conventional media to translate architectural ideas limits the exploration and communication of spatial issues. Designing within and understanding a three-dimensional space, MR offers new opportunities to designers. This relates to similar findings of design studios carried out within IVE (Schnabel, 2002).

Seichter and Schnabel (2005) used the ready available MR-technologies to conduct a design studio as base of their research. They studied how designers create and communicate early design ideas by employing MR as a medium for their interactions. Subsequently the study assessed the perception and understanding of the design process within a collaborative design studio that employed MR as a design medium. They examined the relative effectiveness of the MR instruments in enabling the communication between real and virtual representations.

MR offers a platform for teamwork in remote settings. Architects can collaborate with colleagues using an interactive media that supports the design and communication process in a more immediate way than simply the exchange of files. Communication is enhanced through media that relate to the process of thinking, creating and understanding. The MR-studio demonstrates the ability to establish a unique combination of collaboration and communication of an interactive design process that is transparent and immediate. Users of an MR system are more highly supported to investigate spatial relationships and characteristics of the design can be experienced dynamically within the real and virtual environment.

Using an MR system, designers gain a more complex understanding of relationships of their design and engage in a richer communication with their partners about their design proposals. MR contributes to architectural design through an innovative approach thus enabling new forms of design expression.

These findings support Kvan's (2004) postulation that new opportunities arise for architectural design as we move apart by utilising digital tools to reconnect. The sharing of ideas is related to the perceivable and tangible existence of design items as well as the sensation of applying them.

5. New Dimensions

The above discussed MR and IVE Design Studios (Seichter and Schnabel, 2005; Schnabel, 2002), addressed concepts of architectural design creation influencing recent developments in architectural design education. These partly experimental, partly realistic studios explored innovative methods of architectural expression, form finding and communication and developed unconventional solutions. They coupled the studio-learning environment with an in-depth digital media exploration in order to close the gap between skill training and the application of that knowledge to explore new ways to integrate compound design issues. The use of MR and VR as design instruments allowed the participants to create an innovative architectural design language, based on 3D experiences of space with real and virtual descriptions.

MR-design studios rely on the skills and knowledge of the participants. Often however, these skills have to be built up first. For this reason, the training has to be part of the studio and be directly related to the design intentions of the studio. In the above mentioned studios the students acquired most of their software skills and experience of MR design methods within the first half of the semester. Empowered by their new gained skills the students connected their knowledge with their ambition to express their design proposals using a new design language. This amplified their design experience and learning outcomes.

The studio also merged the individual projects into one larger unit and students shared knowledge and skills. This removed them from an individual ownership of their design but allowed them to reflect on their own as well their colleagues' design as a whole cluster of contributions. This relates to earlier research of design studios that were based on the same principle where media were applied outside their normal pre-described purpose, and innovative design methods were deployed through an interplay of media and design explorations (Kvan, 2000; Schnabel *et al.*, 2004).

6. Interplay in Learning

Architectural design studios are an essential learning experience for students. Their traditions and proceedings are well established. These studios are, additionally, informed and supplemented by courses and seminars, which can feed into their learning outcomes. Studios go beyond pure skill training and require reflection upon, and the creation of, knowledge. There can be, however, a gap between skills training and the application of knowledge within the studio context. At the final presentation of their work, students may not be able to identify how they arrived at their solution and to what extent individual contributors informed their design.

This tension is also apparent in design studios that relay on digital media. These studios present the underlying concepts of architectural design using digital communication tools, but also have to provide training in software skills and other technical subjects (Kvan, 2004). The integration of digital media courses into design studio curricula often fails, because the compound acquisition of skills prevents a deep exploration of design and the theoretical aspects involved. Participants can employ digital media tools within a studio context only long after they have mastered the subject matter and acquired proficiency in techniques. By then, however, the studio may consider these skills no longer valid.

A dilemma of semester-based teaching is that students reach their highest level of skill and experience at the end of a term, after which they leave for their break. Students are therefore unable to apply their knowledge immediately. At the beginning of the following term, however, the knowledge and skills they had gained earlier are likely to be either inactive or not employed, and learning foci may have shifted to other aims.

The architectural design studios presented here addressed these issues by integrating the learning experience from the beginning by focusing on interplays of instruments, media and realms that create or inform about the design. The objective of this 'interplay-designing' was to allow participants to understand the impact that each step and variable has on the design and to follow the impact it has on the project. Participants developed and communicated their understanding of architectural design by utilising their training within the various MRs of the design-studio environment. Because of this, students began to think about design problems in different ways.

The employment of MR allowed students to experience aspects of the design process spatially (in three dimensions), and in detail. Additionally, the overall scale of the design could be communicated using tangible interfaces, digitally controlled devices, physical and digital models, text and sketches. Subsequently the generated design could be linked in a variety of ways to extract or generate new or novel architectural design or understandings of space and form. Additionally, the digital components of the MR could be used in the manufacture of objects for example by means of digitally controlled devices (Seichter and Schnabel, 2005).

Each of the elements created or used in the MR were an essential part of the overall process of design creation. It addresses and expresses certain aspects of the process and its re-presentation. This method enabled a holistic discussion about design, form, function and development, which is significant not only within architectural education, but also in all other dialogues involving spatial representations.

MR-design studios have demonstrated that the problems of MRs are not insurmountable, because technical solutions are constantly evolving, difficulties are resolved and equipment is becoming more sophisticated and easier to use. This is despite the challenges of visual perception, mental workload, errors, comprehension of design and its communication and the different frequency of creation-feedback-modification loops. Since MRs increasingly play a role in the design and form finding of architectural creations, virtuality becomes, in that sense, reality. Following the findings of Gibson and Kvan (2002), this suggests that techniques that produce physical representations on demand, such as rapid prototyping (RP) may have a significant contribution to make to a design process that involves MRs. In educational contexts, whereby the training and learning of spatial aspects and the transfer of knowledge to new situations is crucial, MR allows for a deep learning experience that is authentic and enriched by the experience of direct ‘cause and effect’ on design decisions. Despite the dependencies on technologies, students are embracing new ways of designing and its communication, thereby bringing forward the development of design.

Acknowledgements

The author wishes to thank Aimee Tang for her invaluable contributions, critique, and encouragement. Appreciation goes to students of the Master of Architecture programmes at the Universities of Hong Kong and Sydney, who never gave up their esprit and patience exploring new horizons in their studies. The research was supported by both Universities.

DEBATING OPPORTUNITIES: LEARNING DESIGN THROUGH DIFFERENT STRUCTURES

THOMAS KVAN
University of Melbourne, Australia

Abstract. What is the role of representation and simulation in the design process and how do these support learning in the design studio? How can augmented reality assist such learning? To examine this, the chapter will first consider how modes of representation have been used in architectural design, with particular focus on the role of models, then consider how these modes affect the design student's understanding of their work, and finally postulate the manner in which augmented reality contributes to this process.

Keywords. Structure, Framing, Cognition.

1. What You Represent Is What You See

We represent designs and design ideas for a number of purposes including exploring the opportunity of a design situation, communicating a proposal or documenting a position for later review. In the act of learning, such representations also serve to assist the student to understand the potential of their propositions, including aspects not immediately apparent to the student. Can augmented reality environments assist learning and, if so, in what manner do they contribute? To start, I will review the role of representations, examine two and three dimensional modes of representation and note how these assist in learning.

The act of designing requires the designer to engage representations of the designed object, these acting as virtual worlds in which designers can experiment at relatively low risk and cost (Schön, 1988). These representations embed ideas, knowledge and reasoning (that is, design decisions) through sketches, drawings, physical models, digital models, or mathematical models. Representations thus play a significant role in the design process: as a mode of

X. Wang and M.A. Schnabel (eds.), Mixed Reality in Architecture, Design and Construction, 227–234.

conversation, communication or documentation, whilst the content, formal structures or modes of presentation are aligned with the purpose of the representation.

In architectural design, the range of representations employed at any time may vary. The introduction of a new technology facilitates the development of new modes of representation. We have seen the introduction of digital media change the manner in which designs can be engaged, explored and communicated. This has happened before; the development of paper in the sixteenth century led to drawings as a mode of representation gaining dominance over physical solid objects as the preferred mode of design activity, bringing with it a change that took design to an intellectual plane of engagement (Wigley, 2001).

Paper-based drawings freed the architect from working on site at a scale of 1:1 (Robbins, 1994). As mentioned, from the development of paper in the late sixteenth century, emerged the tradition of drawing as the act of designing. Through this transition, design itself moved from the production of solid objects to become an intellectual discipline engaged in the plane of paper (Wigley, 2001). The medium of paper is convenient for conveying ideas rapidly, is robust and persistent, facilitating the exploration of alternatives by means of easy editing.

In recent years, the introduction of digital tools has raised concerns that the relationship of the designer to the design is changing. The tangibility of paper and the direct manner in which the user controls its physical presence supports work in ways not found in the digital desktop. It has been suggested that paper-based drawing supports collaborative design activities (Robbins, 1994) and further observed that working on paper affords such collaborative activities better than digital modes of working (Dunlop, 2001; Sellen and Harpe, 2001). On close inspection, these claims are difficult to support. Studies of design collaboration in digital environments suggest that these tools enable and support broader exploration of design alternatives and potentially better understanding of the design proposal (Kvan and Gao, 2006). What, then, prevents their more effective use in design learning?

2. Seeing Beyond the Page

One answer may be in the manner in which we engage digital media in design representation. The focus has been on replacing the role of paper by means of databases of geometric descriptions. This has then led to a planar focus in representation and an exploitation of digital editing. Most prominently, this shift has manifested itself through a diminishment of physical constraints in form seeking which have led to a corresponding shift in form making. This supports an investigation of form for its own ends and, with a planar prominence, engages the viewer without obligation to issues of making or habitation.

Essential to moving design beyond form for its own sake has been the development of readily accessible techniques to connect imaging and producing, that is, translating the image on the screen into a tangible object (Kolarevic, 2003). For many years digital forms have remained confined pixilated and vectorised, denied easy translation into physical form. With the introduction of lower cost and faster rapid prototyping machines, for example, the premise is that we have broken through this barrier and can at last reconnect the tangible with digitally ethereal. Some have gone further and claimed a reinvigoration of the craft traditions as a results of the connection in digital technologies from design to manufacture, postulating the realisation of the 'digital craftsman'(McCullough, 1996). The distinction between the form and the making of the form is, however, still profound and prevents essential contributions of models from being realised.

Drawings and models can both be used to represent the three-dimensionality of structures from different angles. Unlike physical models, however, drawings control the observer in the angle or focus of attention, directing these to significant elements. It is necessary that a view be trained to comprehend certain drawings, such as the standard geometric projections of plan, section and elevation, while other drawings may only be comprehensible to their creator (Goldschmidt, 1991). As has been recognised, drawings have their limits no matter how elegant. As one prominent practitioner has noted, "No three-dimensional drawing, however accurate or sophisticated is a substitute for three-dimensional physical models" (Spencer de Gray, in Robbins, 1994:82). Indeed the accuracy of a drawing can be distracting and fail to communicate spatial properties, as a comparison of a computer generated perspective with the spatial immediacy in a painting by Turner or the richness of spatial engagement created by Escher.

Architects, and architecture students, employ models for a variety of reasons. Early in a design cycle, sketch or study models will be created to examine particular aspects of a design idea. Such models are often assembled rapidly and crudely for it is the immediacy of the feedback sought. According to data these rough models assist students in their design thinking process in many ways shaping their decisions: help them perceive their 3D imagery easily and clearly, explore different forms, understand relative scales with the existing context. Janke (1978) and Ratensky (1983) suggest that the primary use of models is to develop spatial thinking and explore certain aspects of design, particularly massing. We see this in the work of professionals as well, notably Gehry's use of rough paper models from which digital models are then created (van Bruggen, 1998).

Simulation of a wider range of sensorial interfaces should not be ignored, including sensations that are easier to simulate, like 3D aural and dynamic physical properties (Seichter and Kvan, 2004). While physical models have an essential purpose in the design process, digital models offer greater potential. Digital design produces a virtualisation by removing real world physical properties, such as weight, friction et cetera, rather than using existing tools to create a sensual experience of space. More commonly, these design systems strive to represent visual appearance like shape and colour, failing to simulate other sensations like the coldness of stones, the velvet touch of polished bare wood, not to speak of the smell of mown grass. Most influential of all, of course, is the sensation of other people occupying the space concurrently, generating signals of presence through motion, touch, noise (including speech) and smell. While it may be argued that many of these sensations are difficult to simulate, this does not diminish their importance or impact in design, especially when communicated to those who are less trained or equipped to understand tacit information in a model.

3. Seeing in Different Ways

Appropriate problem representations can aid the identification of creative solutions yet some representations will be more effective than others (Holyoak, 1984; Mumford *et al.*, 1994). Goldschmidt (1997) notes that re-representations of design concepts and solutions are essential to successful design. Such representation allows for concepts to be examined, compared, joined, transformed or interpreted. In studio learning terms, this implies that representation of a design through multiple modes potentially supports better understanding of the design problem at hand.

A designer using paper or cardboard will sketch, draft, annotate, tear, glue or erase. Few digital design tools permit a wide range of representational techniques. A drafting system permits precise and formal presentation in which ambiguity is not accommodated; a paint program allows freeform lines but does not support easy transition to a drafted image. The palette of representational techniques within a particular environment is limited and some techniques noticeably absent.

In recent years, there has been a return to the discussion of diagrams in architectural discourse (Pai, 2002). Particular types of diagram have been prominent in digital design tools, namely those diagrams that provide formal abstracted representations of facets of design behaviour or attributes, such as bubble diagrams, tree diagrams, calculations and technical abstractions. These are useful as they focus attention to particular issues, simplifying the complexity of the real situation to a range of issues. As such, they are better representations for some design subtasks. Computational devices have lent themselves to such formal abstractions and have thus these diagrams have found favour in digital design methods.

Diagrams extend beyond such formal abstractions, however. The diagrams engaged by designers extend beyond such formalised representational methods to more freeform or idiosyncratic representations. In these abstractions, the diagrams engage the designer in an active process of development of the ideas, providing the essential backtalk that the *conversation* to which Schön refers (Goldschmidt, 2006; Herr and Karakiewicz, 2007). Digital design environments are particularly poor at supporting such design backtalk, reducing too rapidly design intent to geometric certainty.

Frequently stated but often overlooked, digital interfaces should not seek to replicate analogue processes. An interface need not replace reality but can augment it. If we take a look at research in the field of medicine or geography (Kaufman *et al.*, 1997; Shelton and Hedley, 2002) it becomes very clear that a simulation through augmentation has its advantages especially in accessibility, decision-making, learning and primarily sharing (Arias *et al.*, 2000).

In the realm of learning, it is in conversational models and structure representation that augmented reality has its potential. As mentioned above, simulations extend beyond representation by assuming facets of the behaviour of particular potential designs. Diagrammatic representations, in which particular understandings are abstracted, allow the designer to examine properties obscured by other formal representations.

4. Learning Through Structural Re-presentation

For students to engage in learning, it is essential to expose them to new concepts, illuminate potentialities and provide them with new capacities to understand these novel situations. In design learning, engagement in new strategies is an essential contribution to this learning. It is clear that different design strategies will affect the process and hence the product (Eisentraut, 1999). Some strategies are more appropriate for a particular situation than other strategies. Some strategies can be replicated in different circumstances but none are universally successful, requiring the approach to be adapted to the problem at hand. Designing is more than simply applying preconceived or catalogue routines, making the learning and teaching of design a challenging task.

In fostering the expertise that makes a designer, we need to expose students to the contingency of the solutions as well as the techniques by which they are achieved. Thus, learning about design will be more successful if the student can be made aware of their design strategies and can develop techniques of working that support exploration.

Such strategies can be formed by the media used. After mastering a limited range of techniques, it is not uncommon for us to observe students limiting themselves to the comforts of such mastery. In a traditional design studio, students might work with familiar media, such as pencil on tracing or models from cardboard, reflecting perhaps the favoured techniques of the tutor. With the introduction of digital tools, it is not uncommon to see a student working solely in these media, drafting and modelling a project solely on the display screen thus avoiding the need to gain control over manual methods of representation.

The extent to which such structural activities are of assistance to a student appears to vary, but the apparent connection between undertaking such activities and success is strong. This is due to wording or other issues such as attitude, which can affect people's understanding of them and led to the construction of different problem representations.

We have observed the importance of representational techniques in a longitudinal analysis of design learning (Kvan *et al.*, 2003). In a study of two design studio groups over a full semester, we observed the work of students and their interactions with tutors. In the course of this work, students used pencil and paper to produce plans, sections, elevations as well as diagrams. Some students also used physical models, mathematical calculations, annotations, textual narratives and diagrams of all types. One group of students was active in using digital tools including bulletin boards to which they posted textual and graphic material.

An analysis of their working methods, correlated to their final studio grades, found that students who engaged in positing and presenting ideas using a range of media and representational techniques manifested a deeper understanding of their own work and earned higher grades as a consequence. Successful students engaged in significantly more structural activity than those who were less successful. These re-representations appear to aid the students in seeking better solutions, as suggested by Mumford *et al.* (1994). This supports the proposals by Newell and Simon (1972) and Eisentraut (1999) that certain representations are more favourable for problem perception and solution seeking. The representation of the interim problem formulations can be successfully supported by sketches, diagrams and textual expressions. The act of re-presenting their ideas throughout the semester, structuring their understanding from one cognitive mode to another, appeared to support their learning. Seichter (2007) has found similar benefits in the use of augmented reality in urban design learning, using physical models and digital overlays in shared workspaces, finding that the environment supports engagement in learning that is not available in solely digital media.

5. A Place for Augmentation

The usual range of digital design tools used by students does not usually support such re-presentation. Drafting and modelling tools support particular forms of representation, not the integration of multiple modes. It is here that augmented reality has an advantage in the context of learning.

To articulate this potential, let us look again at the manner in which models contribute to designing. Models offer benefits of approachability, tangibility, manipulability and collaborative engagement. For these purposes, models are used at all scales, ranging from town planning to explanation of particular building sub-components and mechanisms. In particular, complex mass-void relationships, spatial or mechanical sequences more easily communicated in models. Digital models have evolved to become sophisticated virtual environments able to be understood by tyros and the untrained eye, capable to communicating invented spaces and forms.

These representations support understandings of the design idea and its development in a variety of ways. de Zeeuw (1979) proposed that models can be classified in two roles: 'models of' and 'models for.' Here, models are distinguished between those that are representational and those that are developed as tools of investigation. A similar distinction has been drawn between semantic and illustrative representations. Young suggests that "Semantic representations represent by being true… Illustrations are not the sort of thing that can be true" (2001, p. 26).

A distinction is made between those representations that can be correlated directly to behaviours of real objects in the world through the application of semantic rules of interpretation and those representations that cause us to reflect on an experience rather than factual condition. Thus, 'models of' might be considered to be *illustrative* models, while 'models for' are *semantic*. In examining this construct, we observe that a third classification can be postulated. While 'models of' might be considered to be illustrative models and 'models for' are semantic a third category, that of 'models with' – that is, models with which we converse, the models of design conversations (Kvan and Thilakaratne, 2003).

It is in relation to this third category of model making and engagement that digital representations have been particularly poor. The focus of developers of digital tools has been initially on the semantic (witness the early focus on digital analysis tools) and, more recently, the illustrative (in particular as manifested by animation systems), ignoring the essential design support of the conversational.

There is potential in the digital environment that should be exploited and augmented reality environments offer us the potential to engage simultaneously with multiple modes of representation. There is yet considerable work to be done in understanding how such environments can be best used to support learning but one lesson that can be learned is that it is the engagement with several modes of representation within a design process that is the essential contribution from augmented reality. By extending the capacity of a conversation to a shared augmented workspace in which design propositions can be restructured in to different representations, a deeper understanding can be supported.

POSTSCRIPT

EPILOGUE

In this book we have presented an overview of the current status of Mixed Reality in Architecture, Design and Construction. Where are we going from here? A possible next step is based on a framework that establishes taxonomies and classification for the development of precise standards for research, professional and industrial applications. This will foster research in these particular fields and transfer knowledge to everyday products and applications.

The novel aspects of MR within our built environment also have a huge impact on our interpersonal engagement with one other. The section 'Mixed Reality in Design Collaboration' highlighted how working together co-located or remotely within a MR becomes as normal as using a telephone. Not only can we work together and share a common point of view, but we can also understand our partners in ways that move us closer to the actual meaning of our matter of interest. Seichter has pointed it out: MR offers a social space that allows participants to interact in a manner akin to reality despite the fact that we are interacting with virtual elements or being only partly immersed in an ME. Communication and coordination between and among designers and engineers becomes efficient with fewer errors. Real-time interaction in remote locations allows for unseen possibilities to emerge between architect, engineer, constructors and clients.

Architectural design is not only aided by the additional layer of information MR offers to designers, but it has a direct impact on the architecture itself. The third chapter 'Mixed Reality in Architecture' discussed changes in the ways that we think about and communicate space, void and solid. Moreover, MR moves into our daily practice by means of hand-held devices and other unobtrusive instruments. In this way, the digital component of the physical built form enriches our experience and interactions within the surrounding realms. As Jules Moloney has suggested in his chapter, designers make informed decisions that are based on information and evaluation.

Today's construction industries are highly technologically advanced and sophisticated. Digital instruments aid engineers and fabricators in profound ways. Efficiency, quality assurance and maintenance are the main drivers behind this innovation. MR gives direct access to information otherwise impossible to obtain. Mobile devices and interactions with people on site or in the office extend the realm of the profession to new possibilities that will take over the industry in every aspect. Dunston and Shin have pointed out ways in which MR provides unique innovative opportunities for the AEC industries.

Innovative methods of communication and collaboration in MR require and present different learning processes and environments in architecture, design and construction. The section 'Mixed Reality in Education and Learning' highlighted the need for research-integrated learning. MR influences the ways in which we learn and understand complex dependencies within design and building processes. Layers of additional information and access to remote sites gives students and educator accessibility to resources we currently can only imagine. Kvan writes succinctly: 'We see beyond the page by merging real and virtual elements into a new realm.' Representation, as we know it in architecture and design, merges with reality into a hybrid that carries greater depth than merely re-representing an idea or design.

We are witnessing a technological explosion of new dimensions. At the intersection between reality and computer-generated information a new world is evolving that influences all aspects of the professions in architecture, design and construction. MR technologies have been moving out of the research-labs into our lives. New instruments that enable designers, engineers, researchers and students alike, empower everybody to contribute to their fields. The outcomes are richer than reality, more knowledgeable than a single library, and even simpler than paper and pencil.

AUTHOR BIOGRAPHIES

Anders Henrysson has a PhD in Media Technology from Linköping University in Sweden during which time he completed pioneering work on mobile phone Augmented Reality, graphics and interaction. He made the first port of the ARToolKit tracking library to Symbian and built novel application around it. He also developed AR Tennis, the world's first collaborative AR application for mobile phones. AR Tennis is a two-player tennis game where phones are used as rackets, and which won the Grand Prix in 2006's International Mobile Gaming Award and has been exhibited around the world. The unique motion-based interaction metaphor of AR Tennis was extended to full 3D interaction, the study of which was the main topic of Anders' PhD. In a collaboration between the HIT Lab NZ and Saatchi & Saatchi, Anders developed the first commercial AR application for advertising, animals popping up of a newspaper in an ad for Wellington Zoo. He has recently joined the HIT Lab NZ Ltd and will continue to explore the commercial potential of mobile phone AR.

Mark Billinghurst is a researcher developing innovative computer interfaces that explore how virtual and real worlds can be merged to enhance face-to-face and remote collaboration. Director of the Human Interface Technology Laboratory (New Zealand) and a research scientist at the HIT Lab (US) in Seattle, he has produced over 140 technical publications and his work has been demonstrated at a wide variety of conferences. He is active in several research areas including Augmented and Virtual Reality, wearable computing and conversational computer interfaces. He has previously worked at ATR Research Labs in Japan, British Telecom's Advanced Perception Unit and the MIT Media Laboratory. One of his research projects, the MagicBook, was winner of the 2001 Discover award for the best Entertainment application.

Remo Burkhard studied architecture and completed a PhD on Knowledge Visualisation, both at the ETH Zurich. He was co-author of the Science City Project and involved in the Strategic Planning Process 2008–2011 of the ETH Zurich. From 2003 until 2007 he was a Project Manager at the University of St. Gallen, where he founded and built up the Competence Center Knowledge Visualization at the Institute for Media and Communications Management. Here he was also responsible for the executive training program. Remo Burkhard is founding partner of vasp datatecture GmbH, a company in the area of Visualising Business Contents. He published various scientific articles, initiated and organised different international research workshop.

Rui Chen is currently pursuing her PhD in the Department of Design Computing at the University of Sydney. She graduated from the Bachelor Computer Engineering with Honours. During that time she also attended the exchange program to Pennsylvania State University in USA for one year. Since then, she has obtained a Master of Design Science (majoring in Design Computing) from the University of Sydney where she has also been a Research Assistant since July 2007. She has been actively involved in a range of funded projects and also tutors both undergraduate and postgraduate classes in the Faculty of Architecture. Her main research investigates techniques of Learning Enhancement of Design Activities using Tangible Augmented Reality systems and she is a published author of refereed international conference papers as well as journal articles.

Jin Won Choi obtained his Masters and Doctoral degrees from the Ohio State University. Immediately after, he continued his academic career as a post-doctoral fellow at the University of California, Berkeley. Before teaching at Yonsei University as Professor of Housing and Interior Design, he had been a lecturer at the Architectural Department at Ajou University from 1996 onwards. He is currently running a research laboratory called Archmedia Smart Space Research Group. His studies have placed great emphasis on the ways in which Information Technology can be integrated into the field of architecture. Other topics include Construction Information Technology, U-Space, U-City, Virtual Reality, Virtual Architecture, Digital Architecture and Scientific Design Methodology in particular.

Phillip S Dunston is an Associate Professor of Construction Engineering and Management in the School of Civil Engineering at Purdue University. He received his BSc, MSc, and PhD degrees from North Carolina State University. His research interests are in the human factors aspects of implementing Augmented Reality in the construction industry. He directs the Advanced Construction Systems Laboratory (ACSyL) at Purdue.

Amin Hammad received a PhD in civil engineering from Nagoya University in 1993. He joined the faculty of CIISE at Concordia University in 2003. Prior to joining Concordia, he held faculty and visiting positions at Nagoya University, Carnegie Mellon University, the University of Pittsburgh and the University of Tokyo. His research interests are in Computer-Aided Engineering with special focus on applications in the fields of Infrastructure and Urban Management Systems. Examples of his research projects include lifecycle management systems of transportation infrastructure, mobile augmented reality support systems for engineering field tasks, and urban simulation systems using virtual reality.

Atsuko Kaga graduated from the Environmental Engineering Department of Osaka University, Japan in 1995, earning a PhD in Engineering. She is engaged in a community development enterprise at an electric railroad company and is an Associate Professor of Osaka University. Her area of research is in environmental design methods and city revitalisation using Information and Communication Technology. Now engaged in practice of design projects, she was the Membership Officer of the Computer Aided Architectural Design and Research in Asia Association from 2004–2008. Atsuko was awarded the Japanese Society for Engineering Education JSEE Research Lecture Meeting Poster Announcement Prize in 2007. A typical digital design work is the Lavender Park Basic Design Editorial Supervision in Kami-cho, Hyogo, Japan.

Mi Jeong Kim obtained her PhD from the Key Centre of Design Computing and Cognition at the University of Sydney. She is currently the Assistant Professor of Housing and Interior Design at Kyung Hee University, Korea. She is also a design researcher with a research focus on ways in which to conceptualise, develop and evaluate novel Augmented Reality, Mixed Reality and Virtual Reality systems to support Design and Design Collaboration, both collocated and remote. Before she joined the faculty at Kyung Hee University, she was a postdoctoral research associate in the Department of Computer Science at the University of California in Berkeley.

Gudrun Klinker studied computer science (Informatics) at the Friedrich-Alexander Universität Erlangen, Universität Hamburg (Diplom) and Carnegie-Mellon University (PhD) in Pittsburgh, USA, focusing on research topics in Computer Vision. In 1989, she joined the Cambridge Research Laboratory of Digital Equipment Corporation in Boston, Massachusetts, working in the visualisation group on the development of a Reusable Tele-Collaborative Data Exploration Environment to analyse and visualise 3D and higher-dimensional data in medical and industrial applications. Since 1995, she has been researching various aspects of the newly emerging concept of Augmented Reality, first at the European Computer-industry Research Center, then at the Fraunhofer Institute for Computer Graphics, and since 2000 at the Technical University of Munich. Here, her research focus is on developing approaches to Ubiquitous Augmented Reality that lend themselves to realistic Industrial Applications.

Thomas Kvan has contributed widely in the Management of Design Practice and Development of Digital Tools in Design, with a focus on Digital Support of Collaborative Design. An active teacher of design studios, he is Dean of the Faculty of Architecture, Building and Planning at the University of Melbourne, having held posts previously as Professor of Architecture and Dean at the University of Hong Kong and University of Sydney. Author of over 100 publications on Design and Design Management, he has practised in architecture in Africa, Europe, Hong Kong and the USA in practices both small and large, was Director of Software Development in a CAD start up in California, and remains a partner in a Management Consultancy serving the community of architecture and engineering professionals. He has consulted for a large number of well known multinational corporations, has been a member of the Board of Directors in Digital Design and Facility Management associations worldwide and serves on the boards of several journals and professional organisations.

Jules Moloney graduated with First Class Honours in Architecture from the University of Auckland in 1986 and completed a Master of Architecture in Urban Design from the University of Westminster in 1990. He practised Architecture in London from 1987 to 1996, working in a range of local and international practices where he developed expertise in Computer Aided design. In 1996, he took up a lectureship at the University of Auckland and was responsible for developing the Digital Architectural Design program. In 2006 he joined the Faculty of Architecture, Building and Planning at the University of Melbourne, where he is currently senior lecturer in digital Architectural Design. In collaboration with colleagues at Melbourne, he is a co-founder of the group Critical Research in Digital Architecture (CRIDA). The group engages in research and teaching projects as vehicles for Critical Inquiry and the development of Innovative Design Practices at various scales supported by Digital Media in Architecture, Urban Design, Landscape and Environmental Design. An overview of research undertaken with the CRIDA group can be obtained from www.crida.net.

Gerhard Schmitt. Since December 2005, Gerhard Schmitt has been Professor of Information Architecture at the Department of Architecture at ETH. Since 1998 he has been the Vice President for Planning and Logistics of ETH Zurich and before that time, he was Professor of CAAD at the same institution where his teaching included CAAD, CAAD Programming, CAAD Practice, and postgraduate seminars. His research focuses on the Development of Intelligent Design Support Systems and the Architectural Design of the Information Territory. His main books are *Architectura et Machina* (Schmitt, 1993), *Architektur mit dem Computer* (Schmitt, 1996), and *Information Architecture* (Schmitt, 1999). From 1984–1988 he was on the Faculty of Architecture at Carnegie Mellon University and from 1993–1994 was a visiting Professor at Harvard University.

Marc Aurel Schnabel is an Architect and Senior Lecturer in Digital Architecture at the Faculty of Architecture, Design and Planning, The University of Sydney. He is leading research and education in the field of Digital Media in Architectural Design. As President of *CAADRIA*, the international Association for Computer Aided Architectural Design Research in Asia, he is affiliated with various professional and scientific committees. He taught and worked in Germany and Hong Kong for over fifteen years and since then has become highly recognised for his work in the areas of Virtual Environments and Design Learning. In Sydney he is engaged in Mixed Reality Design within Data Modelling Research Network and Parametric Design Learning. He publishes extensively in international journals about Novel Perspectives in Digital Architecture and the Communication of Three-Dimensional Space using Innovative Design Methods. He recently curated two Digital Architectural exhibitions, *Disparallel Spaces* at the Tin Sheds gallery and *8448 cubed* at Gaffa Galleries in Sydney. He is currently establishing the *Digital Architecture Research Alliance – DARA* – that brings together researchers who push the boundaries of current digital spatial design.

Hartmut Seichter is a Post-Doctoral Fellow at the Human Computer Interface Laboratory in Christchurch, New Zealand. In 2002 he received a degree as Diplom-Ingenieur from the Bauhaus University Weimar. This early work focused on using Augmented Reality (AR) in Architectural Design, specifically sketching utilising AR. Later he joined the research group organised by Tom Kvan in Hong Kong to undertake a PhD, which he completed in 2006. This work also focused on the Use of Augmented Reality in a Design Environment and inspired the chapter he wrote for this book. In 2005, Hartmut joined the Human Interface Technology Laboratory as Research Associate and later as Post-Doctoral Fellow. His main research interest is the Development of New Augmented Reality Technology for everyday usage, which includes usage in architectural design. Furthermore, fusion with technologies like remote robotic sensory, Multi-Touch Surfaces, Physics Simulation has inspired some of his current research work.

Do Hyoung Shin is a Continuing Lecturer of Construction Engineering and Management in the School of Civil Engineering at Purdue University. He graduated from Korea University in South Korea, with his BSc degree in Civil Engineering and an MSc degree in Structures. He received another M.Sc. degree in the CEM area and his PhD degree from Purdue University. His research interests are in Visualisation, Automation/Robotics, Ubiquitous Computing, and Human Factors in the Construction Industry.

Bruce Hunter Thomas is the Director of the Wearable Computer Laboratory at the University of South Australia, and is an NICTA Fellow, CTO A-Rage Pty Ltd, and visiting Scholar with the Human Interaction Technology Laboratory, University of Washington. He is the inventor of the First Outdoor Augmented Reality Game ARQuake. His has over 130 research publications. His current research interests include: Wearable Computers, User Interfaces, Augmented Reality, Virtual Reality, CSCW, and Tabletop Display Interfaces. Prof. Thomas' academic qualifications include a B.A. in Physics, from George Washington University, an MSc Computer Science from the University of

Virginia with a thesis titled: Pipeline Pyramids in Dynamic Scenes; and a PhD in Computer Science from Flinders University with a thesis titled: Animating Direct Manipulation in Human Computer Interfaces. His experience includes working at the School of Computer and Information Science, University of South Australia since 1990. Prof. Thomas has run his own computer consultancy company and was Computer Scientist at the National Institute of Standards and Technology (a major US government laboratory for the Department of Commerce), and a software engineer for the Computer Sciences Corporation and the General Electric Company.

Marcus Tönnis has studied Physics and Computer Science at the Technical University of Munich, focusing on software engineering and data management. The topic of his diploma thesis covered the development of a Data Management Architecture for Ubiquitous Ad-Hoc Networks and Augmented Reality. He spent a year at the software company *Beck et al.* before joining Professor Klinker's AR research group as a PhD student in 2004. He is now involved in research to develop Novel AR-Supported Driver Assistance Systems for Head-Up Displays in cars.

Xiangyu Wang is a Lecturer in Design Computing at the Faculty of Architecture, Design and Planning, at the University of Sydney. He obtained his PhD degree in Civil Engineering at Purdue University in 2005. Dr Wang's work features highly Interdisciplinary Research across Design, Computer Engineering, Construction, and Human Factors. His specific research interests include Virtual Environments for Design, Human-Computer Interactions, Computer-Supported Cooperative Work, and Construction Automation and Robotics. He has published over 70 refereed articles in a wide range of highly recognised international journals and conferences (ASCE, IEEE, ACM, etc.). He was awarded a US National Science Foundation grant to investigate Skill Development through Virtual Technologies.

GLOSSARY

2D	Two-Dimensional
3D	Three-Dimensional
ADAS	Advanced Driver Assistance System
AEC	Architecture, Engineering, and Construction
AMG	Automated Machine Guidance
APRIL	Augmented Presentation and Interaction Authoring Language
AR	Augmented Reality
AR CAD	Augmented Reality Computer Aided Drawing
AV	Augmented Virtuality
BIM	Building Information Modelling
CAAD	Computer Aided Architecture Design
CAD	Computer Aided Design
CAVE	Cave Automatic Virtual Environment
CNC	Computer Numeric Controlled
DGPS	Differential Global Positioning System
DOF	Degree of Freedom
EXPL	Exploration of Virtual Environment
FOR	Frames of Reference
GIS	Geographic Information System
GPS	Global Positioning System
GPU	Graphic Processing Unit
HCIs	Human Computer Interfaces
HMDs	Head Mounted Displays
HUD	Heads-Up Display
IA	Interface Awareness
IBR	Image-Based Rendering
IG	Image Generator
IPQ	Igroup Presence Questionnaire
ISMAR	International Symposium on Mixed and Augmented Reality
ITQ	Immersive Tendency Questionnaire
IVE	Immersive Virtual Environment
LOD	Level of Detail

MARS	Mobile Augmented Reality Systems
ME	Mixed Environment
MEMS	Micro-Electro-Mechanical Systems
MR	Mixed Reality
NAVE	Non-expensive Automatic Virtual Environment
PC-IG	Personal Computer-based Image Generator
QI	Quality of Immersion
RP	Rapid Prototyping
RTK-GPS	Real-Time Kinematics Global Positioning System
RV	Reality-Virtuality
SDK	Software Development Kit
SP	Spatial Presence
UWB	Ultra Wideband
VDS	Virtual Design Studio
VEs	Virtual Environments
VR	Virtual Reality
VRML	Virtual Reality Modelling Language
VWs	Virtual Worlds
WLAN	Wireless Local Area Network

REFERENCES

2006–2008, *The R Project for Statistical Computing*, Department of Statistics and Mathematics, Wirtschaftsuniversität Wien, Available online: http://www.r-project.org/, February 2, 2008.

Abdeljaoued Y, Marimon D, Ebrahimi T: 2005, Tracking and user interface for mixed reality, *in* O Shreer, P Kauff, T Sikora (eds), *3D Video Communication: Algorithms, Concepts and Real-Time Systems in Human Centred Communication*, Wiley, Hoboken, NJ, pp. 315–332.

Achten HH: 2001, Normative positions in architectural design-deriving and applying design methods, *in* H Penttilä (ed), *Architectural Information Management-19th Conference on Education in Computer Aided Architectural Design in Europe (eCAADe)*, eCAADe, Helsinki, Finland, pp. 263–268.

Aicher O: 1997, *Analogue and Digital: Writings on the Philosophy of Making*, Wiley-VCH, Verlag GmbH, Berlin.

Alberti LB: 1443-1452, *On the Art of Building in Ten Books* (translation by J Rykwert, R Tavernor, N Leach 1988), Cambridge, MA, MIT Press.

Al-Hussein M, Alkass S, Moselhi O: 2005, Optimization algorithm for selection and on site location of mobile cranes, *Journal of Construction Engineering and Management*, **131**(5), pp. 579–590.

Al-Hussein M, Hammad A, Hui W, Zhang C: 2006, Visualizing crane selection and operation in virtual environment, *in*, *Proceedings of the 6th International Conference on Construction: Applications of Virtual Reality (convr:2006)*, Orlando, FL.

Allport A, Styles EA, Hsieh S: 1994, Shifting intentional set: Exploring the dynamic control of tasks, *in* C Umilta, M Moscovitch (eds), *Attention and Performance XV*, Vol. 15, MIT Press, Cambridge, MA, pp. 421–452.

Andel M, Petrovski A, Henrysson A, Ollila M: 2006, Interactive collaborative scene assembly using AR on mobile phones, *in*, *Proceedings of the 16th International Conference on Artificial Reality and Tele-Existence 2006 (ICAT 2006)*, Hangzhou, China, pp. 1008–1017.

Anders P: 2003, Cynergies: Technologies that hybridize physical and cyberspaces, *in*, *Connecting Crossroads of Digital Discourse*, Association for Computer Aided Design In Architecture (ACADIA), Indianapolis, IN, pp. 289–297.

Appleyard D, Lynch K, Myer JR: 1964, *The View from the Road*, MIT Press, Cambridge, MA.

Arias E, Eden H, Fischer G, Gorman A, Scharff E: 2000, Transcending the individual human mind – creating shared understanding through collaborative design, *ACM Transactions on Computer-Human Interaction (TOCHI)*, **7**(1): 84–113.

ARToolKit: 2008, http://artoolkit.sourceforge.net.

Arns L, Cook D, Cruz-Neira C: 1999, The benefits of statistical visualization in an immersive environment, *in*, *Virtual Reality, 1999 Proceedings, IEEE*, Houston, TX, pp. 88–95.

Azuma R: 1997, "A Survey of Augmented Reality" *Presence: Teleoperators and Virtual Environments*, **6**(4), pp. 355–385.

Azuma R: 1999, The challenge of making augmented reality work outdoors, *in* Y Ohta, H Tamura (eds), *Mixed Reality: Merging Real and Virtual Worlds*, Springer, London, pp. 379–390.

Azuma R, Baillot Y, Behringer R, Feiner S, Julier S, MacIntyre B: 2001, Recent advances in augmented reality, *IEEE Computer Graphics and Applications*, **21**(6): 34–47.

Azuma R, Hoff B, Neely H, Sarfarty R: 1999, A motion-stabilized outdoor augmented reality system, *in*, *IEEE Virtual Reality Conference 1999*, Houston, TX, pp. 252–259.

Azuma R: 1997, "A survey of augmented reality", *Presence: Teleoperators and Virtual Environments*, **6**(4): 355–385.

Baertlein H, Carlson B, Eckels R, Lyle S, Wilson S: 2000, A high-performance, high-accuracy RTK GPS machine guidance system, *GPS Solutions*, **3**(3): 4–11.

Baillot Y, Brown D, Julier S: 2001, Authoring of physical models using mobile computers, *in*, *Fifth International Symposium on Wearable Computers*, ISWC, Zurich.

Bandyopadhyay D, Raskar R, Fuchs H: 2001, Dynamic shader lamps: Painting on movable objects, *in*, *Proceedings of IEEE and ACM International Symposium on Augmented Reality*, New York, NY, pp. 207–216.

Barfield W, Rosenberg C, Furness TA: 1995, Situation awareness as a function of frame of reference, computer-graphics eye-point elevation, and geometric field of view, *The International Journal of Aviation Psychology*, **5**(3): 233–256.

Bass L, Kasabach C, Martin R, Siewiorek D, Smailagic A, Stivoric J: 1997, *The Design of a Wearable Computer*, ACM Press, New York.

Baudisch P, Rosenholtz R: 2003, Halo: A technique for visualizing off-screen locations, *in*, *Proceedings of CHI 2003*, Fort Lauderdale, FL, April, pp. 481–488.

Behringer R: 1999, Registration for outdoor augmented reality applications using computer vision for techniques and hybrid sensors, *Proceedings of the IEEE Virtual Reality Conference VR'99*: 244–251.

Behzadan AH, Kamat VR: 2006, Georeferenced registration of construction graphics in mobile outdoor augmented reality, *Journal of Computing in Civil Engineering*, **21**(4), ASCE, pp. 247–258.

Berman M: 1988, The experience of modernity, *in* J Thackara (ed), *Design after Modernism: Beyond the Object*, Thames and Hudson, London, pp. 35–48.

Billinghurst M, Kato H: 1999, Collaborative mixed reality, *in*, *Proceedings of the 1st International Symposium on Mixed Reality (ISMR 99)*, Ohmsha and Springer Verlag, Yokohama, Japan, pp. 261–284.

Billinghurst M, Bowskill J, Jessop M, Morphett J: 1998, A wearable spatial conferencing space, *in*, *2nd International Symposium on Wearable Computers*, IEEE, Pittsburgh, PA, pp. 76–83.

Billinghurst M, Bee S, Bowskill J, Kato H: 1999, Asymmetries in collaborative wearable interfaces, *in*, *Proceedings of the 3rd International Symposium of Wearable Computers*, San Francisco, CA, pp. 133–140.

Billinghurst M, Belcher D, Gupta A, Kiyokawa K: 2003, Communication behaviors in colocated collaborative AR interfaces, *International Journal of Human-Computer Interaction*, **16**(3): 395–423.

Bois Y-A, Shepley J: 1984, A picturesque stroll around Clara-Clara, *October*, **29**: 32–36.

Bonnes B, Secchiaroli G, Montagna G: 1995, *Environmental Psychology: A Psycho-social Introduction*, SAGE, London.

Bosselmann P: 1998, *Representation of Places: Reality and Realism in City Design*, University of California Press, Berkeley, CA.

Bowman DA, Gabbard JL, Hix D: 2002, A survey of usability evaluation in virtual environments: Classification and comparison of methods, *Presence: Teleoperators and Virtual Environments*, **11**(4): 404–424.

Bradford JW, Cheng NY, Kvan T: 1994, Virtual design studios, *in*, *Conference Proceedings: 12th European Conference on Education in Computer Aided Architectural Design (eCAADe): The Virtual Studio*, University of Strathclyde, Glasgow, UK, pp. 7–10.

Broll W, Lindt I, Ohlenburg J, Wittkämper M, Yuan C, Novotny T, Fatah gen. Schieck A, Mottram C, Strothmann A: 2004, ARTHUR: A collaborative augmented environment for architectural design and urban planning, *Journal of Virtual Reality and Broadcasting*, **1**(1): 1–10.

Brookhuis K, de Waard D: 2006, The consequences of automation for driver behaviour and acceptance, *in*, *Proceedings of the International Ergonomics Association (CD-ROM)*, Maastricht, The Netherlands.

Brooks FP: 1986, Walkthrough—a dynamic graphics system for simulating virtual buildings, *in*, *Proceedings of the 1986 Workshop on interactive 3D Graphics* (Chapel Hill, North Carolina, United States). F. Crow and S.M. Pizer, Eds. SI3D '86. ACM, New York, NY, 9-21. DOI = http://doi.acm.org/10.1145/319120.319122

Bubb H: 1993, Systemergonomische Gestaltung, *in* H Schmidtke (ed), *Ergonomie*, Carl Hanser-Verlag, Munich, pp. 305–458.

Bundesen G, Larsen A: 1975, Visual transformation of size, *Journal of Experimental Psychology: Human Perception and Performance*, **1**(3): 214–220.

Burkhard RA, Andrienko G, *et al.*: 2007, Visualization Summit 2007: Ten research goals for 2010, *Information Visualization*, **6**(3): 169–188.

Burry M, Murray Z: 1997, Computer aided architectural design using parametric variation and associative geometry, *in*, *15th ECAADE Conference: Challenges of the future*, Vienna.

Calabrese F, Ratti C: 2007, Real time Rome, *Networks and Communication Studies - Official Journal of the IGU's Geography of Information Society Commission (NETCOM)*, **20**(3–4): 247–257.

Calvino I: 1974, *Invisible Cities* (translated by W Weaver), Harcourt Brace Jovanovich, San Diego, CA.

Campbell DA, Wells M: 1994, *A Critique of Virtual Reality in the Architectural Design Process (HITL Report No R-94-3)*, Human Interface Technology Laboratory (HITL), University of Washington, Seattle, WA, Available online: http://www.hitl.washington.edu/publications/r-94-3/, 23 May 2001.

Castellani U, Livatino S, Fisher RB: 2002, Improving environment modelling by edge occlusion surface completion, *in*, *Proceedings of the First International Symposium on 3D Data Processing Visualization and Transmission*, Padua, Italy, pp. 672–675.

Cave KR, Kosslyn SM: 1989, Varieties of size-specific visual selection, *Journal of Experimental Psychology: General*, **118**(2): 148–164.

Chen C-T, Chang T-W: 2006, Spatially augmented reality design environment, *in* JP Van Leeuwen, HJP Timmermans (eds), *Innovations in Design & Decision Support Systems in Architecture and Urban Planning*, Springer, Dordrecht, Netherlands, pp. 487–499.

Chen YZ, Frame I, Maver TW: 1998, A virtual studio environment for design integration, *Advances in Engineering Software*, **29**(10): 787–800.

Clarke J, Vines J, Mark E: 2003, An augmented virtuality scientific data center, *in*, *Proceedings of the 2003 User Group Conference*, pp. 354–357.

Coelho EM, MacIntyre B, Julier SJ: 2004, OSGAR: A Scene Graph with Uncertain Transformations, *International Symposium on Mixed and Augmented Reality (ISMAR04)*, 6–15.

Conole G, Dyke M: 2004, What are the affordances of information and communication technologies, *ALT-J, Research in Learning Technology*, **12**(2): 113–124.

Cooper LA, Podgorny P: 1976, Mental transformations and visual comparison processes: Effects of complexity and similarity, *Journal of Experimental Psychology: Human Perception and Performance*, **2**(4): 503–514.

Cowart M: 2006, *Embodied Cognition (The Internet Encyclopedia of Philosophy)*, Available online: http://www.iep.utm.edu/e/embodcog.htm, October 2007.

Cullen G: 1961, *The Concise Townscape*, Van Nostrand Reinhold, New York.

Curtis D, Mizell D, Gruenbaum P, Janin A: 1998, Several devils in the details: Making an AR application work in the airplane factory, *in, Proceedings of the international Workshop on Augmented Reality: Placing Artificial Objects in Real Scenes* (Bellevue, Washington, United States). R. Behringer, G. Klinker, D.W. Mizell, Eds. A.K. Peters Ltd., Natick, MA, 47–60.

Daruwala Y: 2004, *3DT: Tangible Input Techniques used for 3D Design and Visualization (Honours thesis)*, BSc Arch Honours, Faculty of Architecture, Design and Planning, The University of Sydney, New South Wales.

Dave B: 1995, Towards distributed computer-aided design environments, *in, The Global Design Studio: Proceedings of the Sixth International Conference on CAAD Futures*, Centre for Advanced Studies in Architecture, National University of Singapore, Singapore, pp. 659–666.

Dave B: 2003, Hybrid spaces of practice, *in, Digital Design: Research and Practice: 10th International Conference on Computer Aided Architectural Design Futures*, Vienna.

Davidson JN, Campbell A: 1996, Collaborative Design in Virtual Space-GreenSpace II: A Shared Environment for Architectural Design Review, *in, Design Computation, Collaboration, Reasoning, Pedagogy: Proceedings of the ACADIA 1996 Conference, Arizona, October 31-November 2, 1996*, University of Arizona, Tucson, AZ, pp. 165–179.

de Zeeuw G: 1979, Onderzoek in verandering, *in, Rede als Richtsnoer*, Mouton Uitgevers, The Hague, pp. 335–353.

Dennis AR, Kinney ST, Hung YTC: 1999, Gender differences in the effects of media richness, *Small Group Research*, **30**(4): 405–437.

Dias JMS, Santos P, Bastos L, Monteiro L, Silvestre R, Diniz N: 2002, MIXDesign, tangible mixed reality for architectural design, *in, 1st Ibero-American Symposium in Computer Graphics (SIACG)*, Guimarães, Portugal.

Do EY-L, Gross MD: 2001, Thinking with diagrams in architectural design, *Artificial Intelligence Review*, **15**(1): 135–149.

Dunlop S: 2001, *A Dictionary of Weather*, Oxford University Press, Oxford.

Dunston P, Wang X, Billinghusrt M, Hampson B: 2002, "Mixed reality benefits for design perception" *In Proceedings of the 19th International Symposium on Automation and Robotics in Construction (ISARC 2002)*, Sept. 23rd–25th, Washington, DC.

Dunston PS, Sinfield JV, Shin D: 2007, Spatial tracking challenge for Augmented Reality on building construction sites, *in, Proceedings of the Fourth International Structural Engineering and Construction Conference (ISEC-4)*, Taylor & Francis, Melbourne, Australia, pp. 1247–1251.

Dunston PS, Wang X: 2005, Mixed reality-based visualization interfaces for architecture, engineering, and construction industry, *Industry Journal of Construction Engineering and Management*, **131**(12): 1301–1309.

Eastman CM: 1999, *Building Product Models: Computer Environments Supporting Design and Construction*, CRC Press, Boca Raton, FL.

Echeverri AA: 2005, Time-based computer-aided architectural research for mapping techniques in multicultural space re-definitions, *in, International Conference of the Association for Computer-Aided Architectural Design Research in Asia (CAADRIA)*, New Delhi.

Eisenstein S: 1949, *Film Form: Essays in Film Theory* (translated by J Leyda), edited by J Leyda, Harcourt, Brace and World, New York.

Eisentraut R: 1999, Styles of problem solving and their influence on the design process, *Design Studies*, **20**(5): 431–437.

Elements Interactive B.V.: 2005/2008, *Edglib*, Available online: http://www.edgelib.com/, April 2008.

Evans R: 1995, *The Projective Cast: Architecture and Its Three Geometries*, MIT Press, Cambridge, MA.

Falk J, Björk S: 1997, The BubbleBadge: A public wearable display, *in, Conference on Human Factors in Computing Systems (CHI'99)*, ACM Press, New York, Pittsburgh, PA, pp. 318–319

Falk J, Redström J, Björk S: 1999, Amplifying reality, *in, 1st International Symposium on Handheld and Ubiquitous Computing*, Springer, Karlsruhe, Germany, pp. 274–280.

Feiner S, Macintyre B, Seligmann D: 1993, Knowledge-based augmented reality, *Communications of the ACM: Special Issue on Computer Augmented Environments: Back to the Real World*, **36**(7): 53–62.

Feiner S, Webster A, Krueger T, MacIntyre B, Keller E: 1995, Architectural anatomy, *Presence*, **4**(3): 318–325.

Feiner S, MacIntyre B, Höllerer T, Webster A: 1997, A touring machine: Prototyping 3D mobile augmented reality systems for exploring the urban environment, *in, 1st International Symposium on Wearable Computers*, Cambridge, MA, pp. 74–81.

Fitzmaurice GW, Buxton W: 1997, An empirical evaluation of graspable user interfaces: Towards specialized, space-multiplexed input, *in, Proceedings of the ACM CHI 1997 Conference on Human Factors in Computing Systems*, ACM Press, Atlanta, GA, pp. 43–50.

Fitzmaurice GW, Ishii H, Buxton WAS: 1995, Bricks: Laying the foundations for graspable user interfaces, *in, Proceedings of the SIGCHI Conference on Human Factors in Computing Systems*, ACM, Denver, CO, pp. 442–449.

Fjeld M, Bichsel M, Rauterberg M: 1998, BUILD-IT: An intuitive design tool cased on direct object manipulation, *in, Proceedings of the International Gesture Workshop: Gesture and Sign Language in Human-Computer Interaction*, Springer, Bielefeld, Germany, pp. 297–308.

Freeman J, Avons SE, Pearson DE, Ijsselsteijn WA: 1999, Effects of sensory information and prior experience on direct subjective ratings of presence, *Presence: Teleoperators and Virtual Environments*, **8**(1): 1–13.

Fuchs H, Neuman U: 1993, A vision: Telepresence for medical consultation and other applications, *in, Sixth International Symposium of Robotics Research*, Hidden Valley, PA, pp. 555–571.

Fuchs H, Meyer AA, Livingston MA, Raskar R, Colucci D, Keller K, Crawford JR, Rademacher P, Drake SH: 1998, Augmented reality visualization for laparoscopic surgery, *in* WM Wells, AC Colchester, S Delp (eds), *Proceedings of the First International Conference on Medical Image Computing and Computer-Assisted Intervention*, Springer, Heidelberg, Germany, pp. 934–943.

Fukuda T, Kawaguchi M, Yeo W, Kaga A: 2006, Development of the environmental design tool "Tablet MR" on-site by Mobile Mixed Reality Technology, *eCAADe 2006*, Greece, pp. 84–87.

Gabbard JL, Swartz K, Richey K, Hix D: 1999, Usability evaluation techniques: A novel method for assessing the usability of an immersive medical visualization VE, *in, Proceedings of the International Conference on Virtual Worlds and Simulation VWSIM99*, ACM PRESS, San Francisco, CA, pp. 165–170.

Gao S, Kvan T: 2004, An analysis of problem framing in multiple settings, *Design Computing and Cognition*, **2**(4): 444–460.

Gartner Research: 2006, *Hype Cycle for Emerging Technologies: 3 Part Special Report*, Gartner Inc., Available online: http://www.gartner.com/it/docs/reports/asset_154296_2898.jsp.

Geiser G: 1985, Man machine interaction in vehicles, *ATZ: Automobiltechnische Zeitschrift*, **87**: 74–77.

Gerhard M, Moore DJ, Hobbs DJ: 2001, Continuous presence in collaborative virtual environments: Towards a hybrid avatar-agent model for user representation, *in, Proceedings of the Intelligent Virtual Agents (IVA): 3rd International Workshop*, Madrid, Spain, pp. 10–11.

Gero JS: 1999, Representation and reasoning about shapes: Cognitive and computational studies in visual reasoning in design, *in, International Conference on Spatial Information Theory (COSIT): Cognitive and Computational Foundations of Geographic Information Science*, Stade, Germany.

Gibson I, Kvan T: 2002, *The Use of Rapid Prototyping for Architectural Concept Modelling (SME Technical Paper PE02-222)*, The Society of Manufacturing Engineers (SME), Dearborn, MI.

Gibson JJ: 1979, *The Ecological Approach to Visual Perception*, Houghton Mifflin, Boston.

Glanville R: 1999, Researching design and designing research, *Design Issues*, **15**(2): 80–92.

Goldschmidt G: 1991, The dialectics of sketching, *Creativity Research Journal*, **4**(2): 123–143.

Goldschmidt G: 1997, Capturing indeterminism: Representation in the design problem space, *Design Studies*, **18**(4): 441–455.

Goldschmidt G: 2006, The backtalk of self-generated sketches, *Design Issues*, **19**(1): 72–88.

Gortler SJ, Grzeszczuk R, Szeliski R, Cohen MF: 1996, The lumigraph, *in, SIGGRAPH '96: Proceedings of the 23rd Annual Conference on Computer Graphics and Interactive Techniques*, ACM Press, New Orleans, LA, pp. 43–54.

Graz University of Technology: 2006, *AR ToolKitPlus: Handheld Augmented Reality Phase 2*, Graz University of Technology, Available online: http://studierstube.icg.tu-graz.ac.at/handheld_ar/artoolkitplus.php, April 2008.

Grimm P, Haller M, Paelke V, Reinhold S, Reimann C, Zauner R: 2002, AMIRE-authoring mixed reality, *in, Proceedings of the First IEEE International Augmented Reality Toolkit Workshop*, Darmstadt, Germany.

Gruchalla K: 2004, Immersive well-path editing: Investigating the added value of immersion, *in, Virtual Reality, 2004 Proceedings, IEEE*, Boulder, CO, pp. 157–164.

Hack G, Canto M: 1984, Collaboration and context in urban design, *Design Studies*, **5**(3): 178–185.

Hammad A, Garrett JH, Karimi HA: 2002, Potential of mobile augmented reality for infrastructure field tasks, *in Proceedings of Applications of Advanced Technology in Transportation Conference (AATT' 2002)*, Cambridge, MA, pp. 456–472.

Hammad A, Garrett JH, Karimi H: 2004, "Location-based computing for infrastructure field tasks." *Telegeoinformatics: Location-based computing and services*, CRC, Fla., 287–314.

Hannon JJ: 2007, *Emerging technologies for construction delivery, NCHRP Synthesis 372: Emerging Technologies for Construction Delivery, A Synthesis of Highway Practice*, National Cooperative Highway Research Program, Transportation Research Board, Washington, DC.

Hart SG, Staveland LE: 1988, *NASA Task Load Index (TLX): V1.0 Manual*, Human Performance Research Group, NASA Ames Research Center, Moffett Field, CA.

Harvey L, Moloney J: 2005, 'Resounding Cities: Acoustic Ecology and Games Technologies', *in Environmental and Global Citizenship*, Ed. Robert Fisher, Rodopi, Amsterdam.

Heeter C: 1992, Being there: The subjective experience of presence, *Presence: Teleoperators and Virtual Environments*, **1**(2): 262–271.

Hendrickson C, Rehak DR: 1993, The potential of a 'Virtual' construction site for automation planning and analysis, *in, Proceedings of the 10th International Symposium on Automation and Robotics in Construction (ISARC) Conference*, Houston, TX, pp. 511–518.

Henrysson A, Ollila M: 2004, UMAR: Ubiquitous mobile augmented reality, *in, Proceedings of the 3rd International Conference on Mobile and Ubiquitous Multimedia*, College Park, MD, pp. 41–45.

Henrysson A, Billinghurst M, Ollila M: 2005a, Virtual object manipulation using a mobile phone, *in, Proceedings of the 2005 International Conference on Augmented Tele-Existence*, Christchurch, New Zealand, pp. 164–171.

Henrysson A, Ollila M, Billinghurst M: 2005b, Mobile phone based AR scene assembly, *in, Proceedings of the 4th International Conference on Mobile and Ubiquitous Multimedia*, The University of Canterbury, Christchurch, New Zealand, pp. 95–102.

Henrysson A, Billinghurst M, Ollila M: 2005c. Face to Face Collaborative AR on Mobile Phones. *International Symposium on Augmented and Mixed Reality (ISAMR'05)*, pp. 80–89.

Herbert DM: 1993, *Architectural Study Drawings*, Wiley, New York.

Herr CM, Karakiewicz J: 2007, ALGOGRAM: Automated diagrams for an architectural design studio, *in, 12th International Conference on Computer Aided Architectural Design Futures (CAAD Futures)*, Sydney, pp. 167–180.

Hinckley K, Pausch R, Proffitt D, Patten J, Kassell N: 1997, Cooperative bimanual action, *in, Proceedings of the ACM CHI 1997 Conference on Human Factors in Computing Systems*, ACM Press, Atlanta, GA, pp. 27–34.

Hirschberg U, Schmitt G, Kurmann D, Kolarevic B, Johnson B, Donath D: 1999, The 24 hour design cycle: An experiment in design collaboration over the internet, *in, Proceedings of The Fourth Conference on Computer Aided Architectural Design Research in Asia (CAADRIA)*, Shanghai, China, pp. 181–190.

Höllerer T, Feiner S, Pavlik J: 1999, Situated documentaries: Embedding multimedia presentations in the real world, *in, Proceedings of the Third International Symposium on Wearable Computers (ISWC'99)*, pp. 79–86.

Holmquist LE, Falk J, Wigström J: 1999, Supporting group collaboration with inter-personal awareness devices, *Journal of Personal Technologies*, **3**(1–2): 13–21.

Holyoak KJ: 1984, Mental models in problem solving, *in* JR Anderson, M Kosslyn (eds), *Tutorials in Learning and Memory: Essays in Honor of Gordon Bower*, Freeman, San Francisco, CA, pp. 193–218.

Hughes CE, Stapleton CB: 2005, The shared imagination: Creative collaboration in augmented virtuality, *in, Proceedings of Human Computer Interaction International 2005 (HCII2005) July 23–27*, Las Vegas, pp. 22–27.

Husserl E: 1931, *Ideas: General Introduction to Pure Phenomenology* (translated by WRB Gibson), Allen & Unwin, London.

Hwang J, Jung J, Kim GJ: 2006, Hand-held virtual reality: A feasibility study, *in, Proceedings of the ACM Symposium on Virtual Reality, Software and Technology*, ACM, Limassol, Cyprus, pp. 356–363.

InterSense: 2002, *InterSense: IS-900 Wide Area Precision Motion Tracker, IS300 and Inertia-Cube2 orientation sensors*, Available online: http://www.isense.com, December 2007.

InterSense: 2008, *Product Manual for use with InertiaCube3 and the InertiaCube3 Processor*, <www.isense.com/support.aspx?id=44>, (January, 2008).

Ioannidis N: 2002, *Archeoguide: Augmented Reality-based Cultural Heritage On-site Guide*, Intracom, Available online: http://archeoguide.intranet.gr/project.htm, April 2008.

Ishii H, Underkoffler J, Chak D, Piper B, Ben-Joseph E, Yeung L, Kanji Z: 2002, Augmented urban planning workbench: Overlaying drawings, physical models and digital simulation, *in, Proceedings of the 10th International Symposium on Mixed and Augmented Reality, (ISMAR 2002)*, ACM, Vienna, Austria, pp. 203–211.

Issa R, Fukai D, Lauderdale G: 2003, A study of 3D and 2D construction drawings acceptance in the field, *in, Proceedings of the 3rd International Conference on Construction: Applications of Virtual Reality (ConVR:2003)*, Virginia Tech, Blacksburg, VA, pp. 48–62.

Ito S: 2005, A Study on informational space structure of wide range area – based on mixed reality, *Journal of Architecture and Planning (Transactions of AIJ)*, **590**(April): 87–94.

Iwatani Y: 1998, Love: Japanese style, *Wired News*, **June 11,** Available online: http://www.wired.com/culture/lifestyle/news/1998/06/12899, June 2, 2008.

Janke R: 1978, *Architectural Models,* Academy Editions, London.

Janssen PHT: 2006, The role of preconceptions in design: some implications for the development of computational design tools, *The International Journal of Architectural Computing*, **3**(4): 449–470.

Janssen P, Krammer J: 2007, Unpublished research report on graphic programming interface, University of Melbourne.

Jersild AT: 1926, Mental set and shift, *Archives of Psychology*, **89**: whole volume.

Jonasson S, Dunston PS, Ahmed K, Hamilton J: 2002, Factors in productivity and unit cost for advanced machine guidance, *Journal of Construction Engineering and Management*, **128**(5): 367–374.

Julier S, Baillot Y, Lanzagorta M, Brown D, Rosenblum L: 2000, BARS: Battlefield augmented reality system, *in*, *NATO Symposium on Information Processing Techniques for Military Systems*, Istanbul, Turkey, pp. 9–11.

Kaga A, Sasada T: 2005, Interactive environmental design using 3D digital city archives, *CAAD Talks 4: Insights of Digital Cities*, ARCHIDATA, Taipei, pp. 211–228.

Kähäri M, Murphy D: 2006/2007, *MARA*, Nokia Research Center, Available online: http://research.nokia.com/research/projects/mara/index.html, April 2008.

Kalman RE: 1960, A new approach to linear filtering and prediction problems, *Transactions of the ASME: Journal of Basic Engineering*, **82**, 35–45.

Kamat VR, Martinez JC: 2001, "Visual simulated construction operations in 3D" *J. Comput. Civ. Eng.*, **15**(4), 329–337.

Kamat VR, Martinez JC: 2005, "Dynamic 3D visualization of articulated construction equipment." *J. Comput. Civ. Eng.*, **19**(4), 356–358.

Kamat, El-Tawil: 2007, Evaluation of augmented reality for rapid assessment of earthquake-induced building damage, *J. of Comput. in Civ. Eng.*, **21**(5), 303–310.

Kanade T: 1991, User viewpoint: Putting the reality into virtual reality, *MasPar News*, **2**(2): 4.

Kanade T, Narayanan PJ, Rander P: 1995, Virtualized reality: Concepts and early results, *in*, *IEEE Workshop on the Representation of Visual Scenes*, Cambridge, MA, pp. 69–76.

Kang SB: 1999, A survey of image-based rendering techniques, *Videometric VI*, **3641**: 2–16.

Kaptelinin V: 1996, Activity theory: Implications for human-computer interaction, *in* B Nardi (ed), *Context and Consciousness*, MIT Press, pp. 103–116.

Karimi HA, Liu X, Liu S, Hammad A: 2004, "GPSLoc: Framework for predicting global positioning system quality of service." *J. Comput. Civ. Eng.*, **18**(3), 196–206.

Katayama A, Uchiyama S, Tamura H, Naemura T, Kaneko M, Harashima H: 1998, A cyber-space creation by mixing ray space data with geometric models, *Systems and Computers in Japan*, **29**(9): 21–31.

Kato H, Billinghurst M, Asano K, Tachibana K: 1999, An augmented reality system and its calibration based on marker tracking, *Transactions of the Virtual Reality Society of Japan*, **4**(4): 607–616.

Kato H, Billinghurst M, Blanding B, May R: 2008, *ARToolKit*, Available online: http://www.hitl.washington.edu/artoolkit/, August 20, 2008.

Kaufman DM, Bell W: 1997, Teaching and assessing clinical skills using virtual reality, in KS Morgan, HM Hoffman, D Stredney, SJ Weghorst (eds.), *Medicine meets virtual reality*, JOS Press, pp. 467–472.

Khronos Group: 2008, *OpenGL ES: The Standard for Embedded Accelerated 3D Graphics*, Available online: http://www.khronos.org/opengles/, April 2008.

Kim B, Kim J, Lee JW: 2007, AR-table system for communication, *in*, *International Conference on Human-Computer Interaction, HCI International*, Beijing, China, pp. 22–27.

Kim MJ: 2007, *The Effects of Tangible User Interfaces on Designers' Spatial Cognition*, PhD thesis, Faculty of Architecture, Design and Planning, The University of Sydney, New South Wales.

Kim T, Biocca F: 1997, Telepresence via Television: Two dimensions of telepresence may have different connections to memory and persuasion. [1], *Journal of Computer-Mediated Communication*, **3**(2).

King GR, Piekarski W, Thomas BH: 2005, ARVino-outdoor augmented reality visualisation of viticulture GIS data, *in, ISMAR 2005 - 4th IEEE and ACM International Symposium on Mixed and Augmented Reality*, IEEE Computer Society, Vienna, Austria, pp. 52–55.

Klauer SG, Neale VL, Dingus TA, Ramsey D, Sudweeks J: 2005, Driver inattention: A contributing factor to crashes and near-crashes, *in, Proceedings of Human Factors and Ergonomics Society 49th Annual Meeting*, Santa Monica, CA, pp. 1922–1926.

Kolarevic B: 2003, *Architecture in the Digital Age: Design and Manufacturing*, Spon Press, London.

Koutamanis A: 2000, Approaches to the integration of CAAD education in the electronic era: Two value systems, *in* A Brown, MBP Knight (eds), *Architectural Computing from Turing to 2000: eCAADe Conference Proceedings of the 17th Conference on Education in Computer Aided Architectural Design in Europe (eCAADe)*, Liverpool, UK, pp. 238–243.

Kraut RE, Gergle D, Fussell SR: 2002, The use of visual information in shared visual spaces: Informing the development of virtual co-presence, *in, Proceedings of the 2002 ACM Conference on Computer Supported Cooperative Work*, ACM, New Orleans, LA, pp. 31–40.

Kuan S, Kvan T: 2005, Supporting objects in voxel-based design environments, *in, Proceedings of the 10th International Conference on Computer Aided Architectural Design Research in Asia (CAADRIA 2005)*, New Delhi, India, pp. 105–113.

Kuo CG, Lin HC, Shen YT, Jeng TS: 2004, Mobile augmented reality for spatial information exploration, *in, Culture, Technology and Architecture: 9th International Conference for Computer-Aided Architectural Design Research in Asia (CAADRIA)*, Yonsei University, Seoul.

Kvan T: 2000, Collaborative design: What is it?, *Automation in Construction*, **9**(4): 409–415.

Kvan T: 2004, Reasons to stop teaching CAAD, *in* ML Chiu (ed), *Digital Design Education*, Garden City Publishing, Taipei, Taiwan, pp. 66–81.

Kvan T, Gao S: 2006, Comparative study of problem framing in multiple settings, *in* JS Gero (ed), *Design Computing and Cognition '06*, Springer, Eindhoven, Netherlands, pp. 245–263.

Kvan T, Thilakaratne R: 2003, Models in the design conversation: Architecture vs engineering, *in* N Clare, S Kaji-O'Grady, S Wollan (eds), *Design + Research, Second International Conference of the Association of Architecture Schools of Australasia*, Melbourne.

Kvan T, Schmitt GN, Maher ML, Cheng N: 2000, Teaching architectural design in virtual studios, *in* R Fruchter, F Peña-Mora, WMK Roddis (eds), *Eight International Conference on Computing in Civil and Building Engineering (ICCCBE-VIII)*, Stanford, pp. 162–169.

Kvan T, Wong JTH, Vera AH: 2003, The contribution of structural activities to successful design, *International Journal of Computer Applications in Technology*, **16**(2/3): 122–126.

Larsen A: 1985, Pattern matching: Effects of size ratio, angular difference in orientation, and familiarity, *Perception and Psychophysics*, **38**(1): 63–68.

Lebling PD, Blank MS, Anderson TA: 1979, Zork: A computerized fantasy simulation game, *IEEE Computer*, **12**(4): 51–59.

Lee H: 2003, Tangible Interaction Design for Cooperative Urban Design System, *Unpublished master's thesis, Korea Advanced Institute of Science and Technology*, Daejeon, Republic of Korea.

Lee CH, Ma YP, Jeng T: 2003, A spatially-aware tangible user interface for computer aided design, *in, Proceedings of the Conference on Human Factors in Computing Systems (CHI'03)*, ACM, Fort Lauderdale, FL, pp. 960–961.

Lee GA, Kim GJ, Billinghurst M: 2005, Immersive authoring: What you eXperience is what you get (WYXIWYG), *Communications of the ACM*, **48**(7): 76–81.

Lee S-G: 2005a, Handheld augmented reality, *Technical Report*, SAIT (Samsung Advanced Institute of Technology) Seoul, South Korea, pp. 10–11.

Lee S-G: 2005b, Industrial augmented reality, *Technical Report*, SAIT (Samsung Advanced Institute of Technology) Seoul, South Korea, pp. 2–23.

Lee S-G: 2006, Tutorial: Recent innovations in user interfaces, *The Journal of Korean Institute of Next Generation Computing*: 35–51.

Lee S-G, Chen T, Kim GJ: 2005, Effects of tactile augmentation and self-body visualization on affective property evaluation of virtual mobile phone designs, *Presence: Teleoperators and Virtual Environments*, **16**(1): 45–64.

Lengyel J: 1998, The convergence of graphics and vision, *Convergence*, **31**(7): 46–53.

Lepetit V, Berger M-O: 2001, A semi-interactive and intuitive tool for outlining objects in video sequences with application to augmented and diminished reality, *in*, *Proceedings of International Symposium on Mixed Reality (ISMAR)*, Yokohama, Japan.

Lessiter J, Freeman J, Keogh E, Davidoff J: 2000, Development of a cross-media presence questionnaire: The ITC-sense of presence questionnaire, *in*, *Proceedings of Presence 2000: The Third International Workshop on Presence*, Delft, The Netherlands.

Levoy M, Hanrahan P: 1996, Light ELD rendering, *in*, *SIGGRAPH '96: Proceedings of the 23rd Annual Conference on Computer Graphics and Interactive Techniques*, ACM Press, New Orleans, LA, pp. 31–42.

Lindl R, Walchshäusl L: 2006, Three-level early fusion for road user detection, *PReVENT Fusion Forum e-Journal*, **1**: 19–24.

Loftus GR, Williams D, Dark VJ: 1979, Short-term memory factors in ground controller/pilot communication, *Human Factors*, **21**(2): 169–181.

Lombard M, Ditton T: 1997, At the heart of it all: The concept of presence, *Journal of Computer-Mediated Communication*, **3**(2), Available online: http://www.ascusc.org/jcmc/vol3/issue2/lombard.html,

Lombard M, Ditton TB, Crane D, Davis B: 2000, Measuring presence: A literature-based approach to the development of a standardized paper-and-pencil instrument, *in*, *Third International Workshop on Presence*, Delft, The Netherlands.

Loomis JM: 1992, Distal attribution and presence, *Presence: Teleoperators and Virtual Environments*, **1**(1): 113–119.

Lou C, Kaga A, Sasada T: 2003, Environmental design with huge landscape in real-time simulation system: Real-time simulation system applied to real project, *Automation in Construction*, **12**(5): 481–485.

Lynch K: 1960, *The Image of the City*, MIT Press, Cambridge, MA.

MacIntyre B, Gandy M, Dow S, Bolter JD: 2004, DART: A Toolkit for Rapid Design Exploration of Augmented Reality Experiences, *User Interface Software and Technology (UIST'04)*, ACM Press.

Maher ML, Simoff SJ, Cicognani A: 2000, *Understanding Virtual Design Studios*, Springer, London.

Maver T: 2002, Predicting the past, remembering the future, *in*, *6th IberoAmerican Congress of Digital Graphics, SIGraDi 2002*, Caracas, Venezuela, pp. 2–3.

Maze J: 2002, Virtual tactility: Working to overcome perceptual and conceptual barriers in the digital design studio, *in*, *Thresholds-Design, Research, Education and Practice, in the Space Between the Physical and the Virtual-2002 Annual Conference of the Association for Computer Aided Design In Architecture (ACADIA)*, Pomona, CA, pp. 325–331.

McCullough M: 1996, *Abstracting Craft: The Practiced Digital Hand*, MIT Press, Cambridge, MA.

Mendez E, Wagner D, Schmalstieg D: 2006, *Vidente: A Glance at AR on the UMPC and Smartphone Platform, ISMAR 2006: Demos*, Available online: http://www.dirkreiners.com/ISMAR06Demos/#Demo_12, April 2008.

Milgram P, Colquhoun H: 1999, A taxonomy of real and virtual world display integration, *in* Y Ohta, H Tamura (eds), *Mixed Reality-Merging Real and Virtual Worlds*, Springer, New York, pp. 5–30.

Milgram P, Kishino F: 1994, A taxonomy of mixed reality visual displays, *IEICE Transactions on Information Systems*, **77**(12): 1321–1329.

Milgram P, Takemura H, Utsumi A, Kishino F: 1994, Augmented reality: A class of displays on the reality-virtuality continuum, *in, Proceedings of Telemanipulator and Telepresence Technologies*, Boston, MA, pp. 282–292.

Mine M, Weber H: 1995, Large models for virtual environments: A review of work by the Architectural Walkthrough Project at UNC, *Presence: Teleoperators and Virtual Environments*, **5**(1): 136–145.

Mitchell W: 1995a, Keynote address: Virtual design studios, *in, CAAD Futures*, Singapore.

Moere AV: 2004, *Infoticles: Immersive Information Visualization using Data-Driven Particles*, Ph.D. thesis, Department of Architecture, Swiss Federal Institute of Technology, Zurich.

Möhring M, Lessig C, Bimber O: 2004, Video see-through AR on consumer cell phones, *in, Proceedings of the Third International Symposium on Augmented and Mixed Reality (ISMAR'04)*, Arlington, VA, pp. 252–253.

Moloney J: 2000, Collapsing the tetrahedron: Architecture with(in) digital machines, *in* T Szrajber (ed), *Computers and History of Art Conference Proceedings (University of Glasgow, 24-25 September)*, Vol. 2, Edinburgh, pp. 289–294.

Moloney J: 2002, StringCVE: Collaborative Virtual Environment software developed from a game engine, *Proceedings of the 20th Conference on Education in Computer Aided Architectural Design in Europe*, Warsaw University of Technology, Warsaw, pp. 522–525.

Moloney J: 2006, Augmented reality visualisation of the built environment to support design decision making, *in, Proceedings of the conference on Information Visualization*, IEEE, London, pp. 687–692.

Moloney J: 2007, Screen Based Augmented Reality for Architectural Design: Two Prototype Systems. *In CAADRIA, 2007 Proceedings of the 12th International Conference on Computer Aided Architectural Design Research in Asia*, Nanjing (China) 19-21 April 2007, pp. 577–584.

Moloney J, Amor R: 2003, String CVE: Advances in a game engine based collaborative virtual environmental for architectural design, *in, Proceedings of the 7th International Conference on Construction Applications of Virtual Reality (CONVR)*, Blacksburg, VA, pp. 156–168.

Mora R, Rivard H, Bédard C: 2006, Computer representation to support conceptual structural design within a building architectural context, *Journal of Computing in Civil Engineering*, **20**(2): 76–87.

Müller P, Wonka P, Haegler S, Ulmer A, Van Gool L: 2006, Procedural modeling of buildings, *ACM Transactions on Graphics (TOG)*, **25**(3): 614–623.

Mumford MD, Reiter-Palmon R, Redmond MR: 1994, Problem construction and cognition: Applying problem representations in ill-defined domains, *in* MA Runco (ed), *Problem Finding, Problem Solving, and Creativity*, Ablex Publishing Company, Norwood, NJ, pp. 3–39.

Nam T-J: 2005, Sketch-based rapid prototyping platform for hardware-software integrated interactive products, *in, CHI '05 Conference on Human Factors in Computing Systems*, ACM, Portland, OR, pp. 1689–1692.

Narzt W, Pomberger G, Ferscha A, Kolb D, Müller RW, Wieghardt J, Hörtner H, Lindinger C: 2003, Pervasive information acquisition for mobile AR-navigation systems, *in, Proceedings of the Fifth IEEE Workshop on Mobile Computing Systems & Applications (WMSCA)*, Monterey, CA, pp. 13–20.

Nash EB, Edwards GW, Thompson JA, Barfield W: 2000, A review of presence and performance in virtual environments, *International Journal of Human-Computer Interaction*, **12**(1): 1–41.

Newell A, Simon HA: 1972, *Human problem solving*, Prentice-Hall, Englewood Cliffs, NJ.

Nguyen THD, Qui TCT, *et al.*: 2005, Real-time 3D human capture system for mixed-reality art and entertainment, *IEEE Transactions on Visualization and Computer Graphics*, **11**(6): 706–721.

Norman DA: 1988, *The Psychology of Everyday Things*, Basic Books, New York.

North MM, North SM, Coble JR: 1998, Virtual reality therapy: An effective treatment for phobias, *Virtual Environments in Clinical Psychology and Neuroscience*, IOS Press, Amsterdam, The Netherlands.

Oloufa AA, Ikeda M, Oda H: 2003, Situational awareness of construction equipment using GPS, wireless and web technologies, *Automation in Construction*, **12**(6): 737–748.

Osfield R, Burns D: 2006, *Open Scene Graph*, OSG Community, Available online: http://www.openscenegraph.org, April 2008.

Oxman RE: 2000, Design media for the cognitive designer, *Automation in Construction*, **9**(4): 337–346.

Pai H: 2002, *The Portfolio and the Diagram: Architecture, Discourse, and Modernity in America*, MIT Press, Cambridge, MA.

Parish YIH, Müller P: 2001, Procedural modeling of cities, *in*, *Proceedings of the 28th Annual Conference on Computer Graphics and Interactive Techniques*, ACM Press, New York, pp. 301–308.

Park JY, Lee JW: 2004, Tangible augmented reality modeling, *in*, *International Conference/ Workshop on Entertainment Computing*, Eidnhoven, The Netherlands, pp. 254–259.

Paul P, Fleig O, Jannin P: 2005, Augmented virtuality based on stereoscopic reconstruction in multimodal image-guided neurosurgery: Methods and performance evaluation, *IEEE Transactions on Medical Imaging*, **24**(11): 1500–1511.

Piekarski W: 2004, *Interactive 3D Modeling in Outdoor Augmented Reality Worlds*, PhD thesis, Wearable Computer Lab, University of South Australia, Adelaide, SA.

Piekarski W, Thomas BH: 2001, Tinmith-Metro: New outdoor techniques for creating city models with an augmented reality wearable computer, *in*, *5th International Symposium on Wearable Computers, (ISWC01)*, Zurich, Switzerland, pp. 31–38.

Piekarski W, Thomas BH: 2002, *ARQuake*: The outdoor augmented reality gaming system, *Communications of the ACM*, **45**(1): 36–38.

Piekarski W, Thomas BH: 2003a, Interactive augmented reality techniques for construction at a distance of 3D geometry, *in*, *Immersive Projection Technology / Eurographics Virtual Environments Conference*, Zurich, pp. 19–28.

Piekarski W, Thomas BH: 2003b, An object-oriented software architecture for 3D mixed reality applications, *in*, *2nd IEEE and ACM International Symposium on Mixed and Augmented Reality*, Tokyo, pp. 247–256.

Piekarski W, Thomas BH: 2003c, Tinmith – A mobile outdoor augmented reality modelling system, *in*, *Symposium on Interactive 3D Graphics*, Monterey, CA.

Rackliffe N: 2005, *An Augmented Virtuality Display for Improving UAV Usability*, *Technical Papers*, The Mitre 10 Corporation, Available online: http://www.mitre.org/work/tech_papers/tech_papers_05/05_1208/05_1208.pdf, August 2006.

Ratensky A: 1983, *Drawing and Modelmaking*, Watson-Guptill Publications, New York.

Rauterberg M, Fjeld M, Krueger H, Bichsel M, Leonhardt U, Meier M: 1997, BUILD-IT: A video-based interaction technique of a planning tool for construction and design, *in* H Miyamoto, S Saito, M Kajiyama, N Koizumi (eds), *Proceedings of Work With Display Units (WWDU'97)*, Takorozawa: NORO Ergonomic Lab, pp. 175–176.

Regenbrecht H, Schubert T, Friedmann F: 1998, Measuring the sense of presence and its relations to fear of heights in virtual environments, *International Journal of Human–Computer Interaction*, 10: pp. 233–250.

Regenbrecht H, Schubert T: 2002, Real and illusory interactions enhance presence in virtual environments, *Presence: Teleoperators and Virtual Environments*, **11**(4): 425–434.

Regenbrecht H, Lum T, Kohler P, Ott C, Wagner M, Wilke W, Mueller E: 2004, Using augmented virtuality for remote collaboration, *Presence: Teleoperators and Virtual Environments*, **13**(3): 338–354.

Reitmayr G, Schmalstieg D: 2001, Mobile collaborative augmented reality, *in*, *Proceedings of the IEEE and ACM International Symposium on Augmented Reality*, Columbia University, New York, pp. 114–123.

Reitmayr G, Schmalstieg D: 2003, Location based applications for mobile augmented reality, *in*, *Proceedings of the Fourth Australian User Interface Conference on User Interfaces*, Australian Computer Society, Sydney, Australia, pp. 65–73.

Reitmayr G, Schmalstieg D: 2004, Collaborative augmented reality for outdoor navigation and information browsing, *in*, *Symposium Location Based Services and TeleCartography*, Vienna, Austria, pp. 31–41.

Reitmayr G, Drummond T: 2006, Going out: Robust model-based tracking for outdoor augmented reality, *in*, *Proceedings of 5th IEEE and ACM International Symposium on Mixed and Augmented Reality (ISMAR 2006)*, Santa Barbara, CA, pp. 109–118.

Reitmayr G, Eade E, Drummond T: 2005, Localisation and interaction for augmented maps, *in*, *IEEE International Symposium on Mixed and Augmented Reality (ISMAR)*, Vienna, Austria, pp. 120–129.

Rioux M, Godin G, Blais F: 1992, Datagraphy: The final frontier in communications, *in*, *International Conference on Three Dimensional Media Technology: 3Dmt '92*, Montréal, Canada.

Robbins E: 1994, *Why Architects Draw*, MIT Press, Cambridge, MA.

Roberts GW, Evans A, Dodson A, Denby B, Cooper S, Hollands R: 2002, The use of augmented reality, GPS, and INS for subsurface data visualization, *in*, *FIG XXII International Congress: TS5.13 Integration of Techniques*, Washington, DC.

Rogers RD, Monsell S: 1995, Costs of a predictable switch between simple cognitive tasks, *Journal of Experimental Psychology: General*, **124**(2): 207–231.

Rokita P: 1996, Generating depth of-field effects in virtual reality applications, *Computer Graphics and Applications, IEEE*, **16**(2): 18–21.

Rowe C, Koetter F: 1978, *Collage City* MIT Press, Cambridge, MA.

Sakagawa Y, Katayama A, Kotake D, Tamura H: 2001a, A hardware ray-space renderer for interactive augmented virtuality, *in*, *The Second International Symposium on Mixed Reality*, Yokohama, Japan, pp. 87–94.

Sakagawa Y, Katayama A, Kotake D, Tamura H: 2001b, The Yokohama Character Museum Cyber-Annex: Photorealistic exhibition of museum artifacts by image-based rendering, *in*, *The Second International Symposium on Mixed Reality*, Yokohama, Japan, pp. 203–204.

Satalich GA: 1995, *Navigation and Wayfinding in Virtual Reality: Finding the Proper Tools and Cues to Enhance Navigational Awareness*, Master of Science in Engineering, University of Washington, Seattle, WA.

Satoh K, Hara K, Anabuki M, Yamamoto H, Tamura H: 2001, TOWNWEAR: An outdoor wearable MR system with high-precision registration, *in*, *2nd International Symposium on Mixed Reality*, Yokohama, pp. 210–211.

Schkolne S: 2003, *The Crystal Method: A Three-Dimensional Interaction*, Available online: http://thecrystalmethod2003.com/, April 2008.

Schloerb DW: 1995, A quantitative measure of telepresence, *Presence: Teleoperators and Virtual Environments*, **4**(1): 64–80.

Schmalstieg D, Fuhrmann A, Hesina G: 2000, Bridging multiple user interface dimensions with augmented reality, *in*, *Proceedings of the IEEE and ACM International Symposium on Augmented Reality (ISAR 2000)*, Munich, Germany, pp. 20–29.

Schmalstieg D, Fuhrmann A, Hesina G, Szalavári Z, Encarnação ML, Gervautz M, Purgathofer W: 2002, The Studierstube augmented reality project, *Presence: Teleoperators and Virtual Environments*, **11**(1): 33–54.

Schmitt G: 1993, *Architectura et machina: Computer Aided Architectural Design und virtuelle Architektur*, Friedrich Vieweg & Sohn.

Schmitt G: 1996, *Architektur mit dem Computer*, Vieweg Verlagsgesellschaft.

Schmitt G: 1999, *Information Architecture; Basis and Future of CAAD*. Basel, Birkhäuser.

Schnabel MA: 2002, Collaborative studio in a virtual environment, *in* RL Kinshuk, K Akahori, R Kemp, T Okamoto, L Henderson, CH Lee (eds), *Learning Communities on the Internet: Pedagogy in Implementation, Proceedings of the International Conference on Computers in Education (ICCE)*, IEEE Computer Society, Auckland, New Zealand, pp. 337–341.

Schnabel MA: 2004, *Architectural Design in Virtual Environments*, Ph.D. thesis, Department of Architecture, The University of Hong Kong, Hong Kong.

Schnabel MA 2005 Interplay of domains: New dimensions of form generations, *in* B Martens, A Brown (eds), *Learning from the Past: A Foundation for the Future*, Öesterreichischer Kunst und Kulturverlag, Vienna, Austria, pp. 11–20.

Schnabel MA, Kvan T: 2002, Design, communication & collaboration in immersive virtual environments, *International Journal of Design Computing*, **4** Special Issue on Designing Virtual Worlds, Available online: http://faculty.arch.usyd.edu.au/kcdc/ijdc/vol04/papers/schnabelFrameset.htm, December 1 2006.

Schnabel MA, Kvan T: 2003, Spatial understanding in immersive virtual environments, *International Journal of Architectural Computing (IJAC)*, **1**(4): 435–448.

Schnabel MA, Kvan T, Kuan SKS, Li W: 2004, 3D Crossover: Exploring Objects digitalisé, *International Journal of Architectural Computing (IJAC)*, **2**(4): 475–490.

Schnabel MA, Wang X, Seichter H, Kvan T: 2007, From virtuality to reality and back, *in*, *Proceedings of the 12th International Association of Societies of Design Research (IASDR)*, Hong Kong, pp. 12–15.

Schnabel MA, Wang X, Seichter H, Kvan T: 2008, Touching the untouchables: Virtual-, augmented- and reality, *in*, *Proceedings of the 13th International Conference on Computer Aided Architectural Design Research in Asia (CAADRIA)*, Chiang Mai, Thailand, pp. 293–299.

Schön DA: 1983, *Educating the Reflective Practitioner*, Harper Collins, New York.

Schön DA: 1988, Designing: Rules, types and worlds, *Design Studies*, **9**(3): 181–190.

Schubert T, Friedmann F, Regenbrecht H: 1999a, Embodied presence in virtual environments, *Visual Representations and Interpretations*, Springer, London, pp. 268–278.

Schubert T, Friedmann F, Regenbrecht HT: 1999b, Decomposing the sense of presence: Factor analytic insights, *in*, *2nd International Workshop on Presence*, University of Essex, UK, pp. 3–23.

Schubert T, Friedmann F, Regenbrecht H: 2001, The experience of presence: Factor analytic insights, *Presence: Teleoperators and Virtual Environments*, **10**(3): 266–281.

Schuemie MJ, van der Straaten P, Krijn M, van der Mast C: 2001, Research on presence in virtual reality: A survey, *CyberPsychology & Behavior*, **4**(2): 183–201.

Seichter H: 2003a, Augmented reality aided design, *International Journal of Architectural Computing*, **1**(4): 449–460.

Seichter H: 2003b, Sketchand+ a collaborative augmented reality sketching application, *in*, *Proceedings of the 8th International Conference on Computer Aided Architectural Design Research in Asia (CAADRIA)*, Bangkok, Thailand, pp. 209–222.

Seichter H: 2007, Augmented reality and tangible interfaces in collaborative urban design, *in* A Dong, A Vande Moere, JS Gero (eds), *12th International Computer Aided Architectural Design Futures (CAAD Futures) Conference*, Springer, Sydney, Australia, pp. 3–16.

Seichter H, Kvan T: 2004, Tangible interfaces in design computing, *in* B Rüdiger, B Tournay, H Ørbaek (eds), *Education and Research in Computer Aided Architectural Design in Europe (eCAADe 22)*, The Royal Danish Academy of Fine Arts, Copenhagen, Denmark, pp. 159–166.

Seichter H, Schnabel MA: 2005, Digital and tangible sensation: An augmented reality urban design studio, *in*, *Tenth International Conference on Computer Aided Architectural Design Research in Asia (CAADRIA)*, New Delhi, India, pp. 193–202.

Sellen AJ, Harper RHR: 2001, *The Myth of the Paperless Office*, MIT Press, Cambridge, MA.

Shelton BE, Hedley NR: 2002, Using augmented reality for teaching earth-sun relationships to undergraduate geography students, *Paper Presented at the First IEEE International Augmented Reality Toolkit Workshop*, Darmstadt, Germany.

Shepard RN, Metzler J: 1971, Mental rotation of three-dimensional objects, *Science*, **171**(3972): 701–703.

Sheridan TB: 1992, Musings on telepresence and virtual presence, *Presence: Teleoperators and Virtual Environments*, **1**(1): 120–126.

Shih NJ, Lin CY, Liau CY: 2004, A 3D information system for the digital preservation of historical architecture, Architecture in the Network Society, *eCAADe 2004*, Copenhagen, pp. 630–637.

Shin DH: 2007, *Strategic Development of AR Systems for Industrial Construction*, Ph.D. thesis, Purdue University, West Lafayette, Indiana.

Shin DH, Dunston PS: 2008, Identification of application areas for Augmented Reality in industrial construction based on technology suitability, *Automation in Construction*, **17**(7), 882–894.

Shin DH, Jung W, Dunston PS: 2007, Large Scale Calibration for Augmented Reality on construction sites, *in, Proceedings of the 7th International Conference on Construction Applications of Virtual Reality*, Penn State University, University Park, PA, pp. 139–146.

Shum HY, Kang SB: 2000, A review of image-based rendering techniques, *IEEE/SPIE Visual Communications and Image Processing (VCIP)*, **2**(10): 2–13.

Siegel J, Bauer M: 1997, A field usability evaluation of a wearable system, *in, 1st International Symposium on Wearable Computers*, Cambridge, MA, pp. 18–22.

Simlog: 2008, <www.simlog.com/index.html>, (January, 2008).

Simsarian K, Åkesson KP: 1997, Windows on the world: An example of Augmented Virtuality, *in, Interfaces 97: Sixth International Conference Man-Machine Interaction*, Montpellier, France, pp. 68–71.

Slater M: 1999, Measuring presence: A response to the Witmer and Singer questionnaire, *Presence: Teleoperators and Virtual Environments*, **8**(5): 560–566.

Slater M, Usoh M: 1993, Presence in immersive virtual environments, *in, Virtual Reality Annual International Symposium, 1993 IEEE*, Seattle, WA, pp. 90–96.

Slater M, Wilbur S: 1997, A framework for immersive virtual environments (FIVE) – Speculations on the role of presence in virtual environments, *Presence: Teleoperators and Virtual Environments*, **6**(6): 603–616.

Slater M, Usoh M, Steed A: 1994, Depth of presence in virtual environments, *Presence: Teleoperators and Virtual Environments*, **3**(2): 130–144.

Slater M, Usoh M, Chrysanthou Y: 1995, The influence of dynamic shadows on presence in immersive virtual environments, *Virtual Environments*, **95**: 8–21.

Smith S, Hart J: 2006, Evaluating distributed cognitive resources for wayfinding in a desktop virtual environment, *in, Proceedings of the 1st IEEE Symposium on 3D User Interfaces*, Alexandria, VA.

SourceForge.net: 1999/2008, *TAP: The Architectural Playground*, SourceForge.net Inc., Available online: http://sourceforge.net/projects/libtap/, April 2 2008.

Stafford A, Piekarski W, Thomas BH: 2006, Implementation of God-like interaction techniques for supporting collaboration between outdoor AR and indoor tabletop users, *in, ISMAR 2006 - 5th IEEE - International Symposium on Mixed and Augmented Reality*, Vienna, Austria, pp. 165–172.

Starner T, Mann S, Rhodes B, Levine J, Healey J, Kirsch D, Picard R, Pentland A: 1997, Augmented reality through wearable computing, *Presence: Teleoperators and Virtual Environments*, **6**(4): 386–398.

State A, Livingston MA, Garrett WF, Hirota G, Whitton MC, Pisano ED, Fuchs H: 1996, Technologies for augmented-reality systems: Realizing ultrasound-guided needle biopsies, *in, Proceedings of the 23rd Annual Conference on Computer Graphics and Interactive Techniques of SIGGRAPH '96*, New Orleans, LA, pp. 439–446.

Stevens B, Jerrams-Smith J: 2001, The Sense of Object-Presence with Projection-Augmented Models, *in, Lecture Notes in Computer Science*, Springer, Glasgow, UK, pp. 194–198.

Stevens B, Jerrams-Smith J, Heathcote D, Callear D: 2002, Putting the virtual into reality: Assessing object-presence with projection-augmented models, *Presence: Teleoperators and Virtual Environments*, **11**(1): 79–92.

Sutherland I: 1968, A Head-Mounted A Three-Dimensional Display, *Fall Joint Computer Conf., Am. Federation of Information Processing Soc. (AFIPS) Conf. Proc. 33*, Thompson Books, Washington, D.C., pp. 757–764.

Tamura H, Yamamoto H, Katayama A: 1999, Steps toward seamless mixed reality, *in* Y Ohta, H Tamura (eds), *Mixed Reality-Merging Real and Virtual Worlds*, Springer, New York, pp. 59–79.

Tang A, Owen C, Biocca F, Mou W: 2003, Comparative effectiveness of augmented reality in object assembly, *in, Proceedings of the SIGCHI conference on Human factors in computing systems*, ACM, Fort Lauderdale, FL, pp. 73–80.

Tang SK, Liu YT, Lin CY, Shih SC, Chang CH, Chiu YC: 2001, The visual harmony between new and old materials in the restoration of historical architecture, *in, Proceedings of 6th the Conference on Computer Aided Architectural Design Research in Asia,* University of Sydney, Sydney, pp. 205–210.

Tang SK, Liu YT, Fan YC, Wu YL, Lu HY, Lim CK, Hung LY, Chen YJ: 2002, How to simulate and realise a disappeared city and city life? – A VR Cave simulation, *Proceedings of the 7th International Conference on Computer Aided Architectural Design Research in Asia (CAADRIA) 18–20 April*, Cyberjaya, Malaysia, pp. 301–308.

Tenmoku R, Nakazato Y, Anabuki A, Kanbara M, Yokoya N: 2004, Nara Palace site navigator: Device-independent human navigation using a networked shared database, *in, Proceedings of the 10th International Conference on Virtual Systems and Multimedia (VSMM) November*, Ogaki City, Japan, pp. 1234–1242.

Tenmoku R, Kanbara M, Yokoya N: 2005, Nara Palace site navigator: Mobile tour guide system using multimedia contents, *in, Proceedings of 1st Digital Content Symposium (DCS2005) May 25–27*, Science Museum, Tokyo, Japan.

Thomas B, Close B, Donoghue J, Squires J, De Bondi P, Morris M, Piekarski W: 2000, ARQuake: An outdoor/indoor augmented reality first person application, *in, The Fourth International Symposium on Wearable Computers*, Atlanta, GA, pp. 139–146.

Thomas BH, Piekarski W: 2003, Outdoor virtual reality, *in, Proceedings of the 1st International Symposium on Information and Communication Technologies*, IEEE, Trinity College, Dublin, Ireland, pp. 226–231.

Thomas BH, Tyerman S, Grimmer K: 1998, Evaluation of text input mechanisms for wearable computers, *Virtual Reality*, **3**(3): 187–199.

Thomas BH, Piekarski W, Gunther B: 1999, Using augmented reality to visualise architecture designs in an outdoor environment, *International Journal of Design Computing: Special Issue on Design Computing on the Net (dcnet'99)*, **2**: 1329–7147.

Thorp EO: 1998, The invention of the first wearable computer, *in, 2nd International Symposium on Wearable Computers*, IEEE, Pittsburgh, PA, pp. 4–8.

Tönnis M, Klinker G: 2006, Effective control of a car driver's attention for visual and acoustic guidance towards the direction of imminent dangers, *in, Proceedings of the Fifth IEEE and ACM International Symposium on Mixed and Augmented Reality (October 22-25)*, Santa Barbara, CA.

Tönnis M, Broy V, Klinker G: 2006, A survey of challenges related to the design of 3D user interfaces for car drivers, *in, Proceedings of the 1st IEEE Symposium on 3D User Interfaces*, Alexandria, VA.

Tönnis M, Lange C, Klinker G: 2007, Visual longitudinal and lateral driving assistance in the head-up display of cars, *in, Proceedings of the Sixth IEEE and ACM International Symposium on Mixed and Augmented Reality*, Nara, Japan.

Trimble: 2005, *Trimble 5700 RTK GPS Receiver*, Available online: http://www.trimble.com, December 2006.

Tschumi B: 1996, *Architecture and Disjunction*, MIT Press, Cambridge, MA.

Ulmer A, Halatsch J, Kunze A, Müller P, Van Gool L, Incorporated P, Zurich S: 2007, Procedural design of urban open spaces, *in, Proceedings of eCAADe 2007*, Frankfurt, Germany.

Underkoffler J, Ishii H: 1999, Urp: A luminous-tangible workbench for urban planning and design, *in, Proceedings of the SIGCHI Conference on Human Factors in Computing Systems - The CHI is the Limit*, Pittsburgh, PA, pp. 386–393.

Usoh M, Catena E, Arman S, Slater M: 2000, Using presence questionnaires in reality, *Presence: Teleoperators and Virtual Environments*, **9**(5): 497–503.

van Bruggen C: 1998, *Frank O. Gehry: Guggenheim Museum, Bilbao*, Guggenheim Museum Publications, New York.

Venturi R, Brown DS, Izenour S: 1972, *Learning from Las Vegas*, MIT Press, Cambridge, MA.

Vera AH, Kvan T, West RL, Lai S: 1998, Expertise, collaboration and bandwidth, *in, Proceedings of the ACM CHI 1998 Conference on Human Factors in Computing Systems*, ACM, pp. 503–510.

Vlahakis V, Ioannidis N, Karigiannis J, Tsotros M, Gounaris M, Stricker D, Gleue T, Dähne P, Almeida L: 2002, Archeoguide: An Augmented Reality guide for archaeological sites IEEE computer graphics and applications, *IEEE Computer Graphics and Applications*, **22**(5): 52–60.

Wagner D: 2007, *Handheld Augmented Reality*, Ph.D., Institute for Computer Graphics and Vision, Graz University of Technology, Graz, Austria.

Wagner D, Barakonyi I: 2003, Augmented reality kanji learning, *in, Proceedings of the Second IEEE and ACM International Symposium on Mixed and Augmented Reality*, Tokyo, Japan, pp. 335–336.

Wang W: 2004, *Human Navigation Performance Using 6 Degree of Freedom Dynamic Viewpoint Tethering in Virtual Environments*, Ph.D. thesis, Faculty of Mechanical and Industrial Engineering, University of Toronto.

Wang X, Dunston PS: 2006, “Potential of augmented reality as an assistant viewer for computer-aided drawing.” *J. Comput. Civ. Eng.*, **20**(4), 437–441.

Wang X, Gong Y: 2007, Specifying augmented virtuality systems for architectural design and collaboration, *in, Proceedings of the 13th International Conference on Virtual Systems and Multimedia (VSMM'07)*, Brisbane, QLD, pp. 23–26.

Wang Y, Jia X, Lee HK: 2003, An indoors wireless positioning system based on wireless local area network infrastructure, *in, Proceedings of the 6th International Symposium on Satellite Navigation (SatNav 2003)*, Melbourne, Australia.

Webster A, Feiner S, MacIntyre B, Massie W, Krueger T: 1996, “Augmented reality in architectural construction, inspection and renovation”. *Proc., Computing in Civil Engineering*, ASCE, Va, pp. 913–919.

Welch RB, Blackmon TT, Liu A, Mellers BA, Stark LW: 1996, The effects of pictorial realism, delay of visual feedback, and observer interactivity on the subjective sense of presence, *Presence: Teleoperators and Virtual Environments*, **5**(3): 263–273.

Wenz F, Hirschberg U: 1997, Phase (x) Memetic engineering for architecture, *in* B Martens, H Linzer, A Voigt (eds), *Proceedings of the 15 theCAADe Conference Challenges of the Future*, Österreichischer Kunst- und Kulturverlag, Vienna, Austria.

Wigley M: 2001, Paper, Scissors, Blur, *in* C de Zegher, M Wigley (eds), *The Activist drawing: Retracing Situationist Architectures from Constant's New Babylon to Beyond*, MIT Press, New York, pp. 27–56.

Witmer BG, Singer MJ: 1998, Measuring presence in virtual environments: A presence questionnaire, *Presence: Teleoperators and Virtual Environments*, **7**(3): 225–240.

Witmer BG, Bailey JH, Knerr BW, Parsons KC: 1996, Virtual spaces and real world places: Transfer of route knowledge, *International Journal of Human-Computer Studies*, **45**(4): 413–428.

Wittkower R: 1975, *Studies in the Italian Baroque*, Thames and Hudson, London.

XML-RPC.com: 2004–2008, *XML-RPC*, UserLand Software, Available online: http://www.xmlrpc.com/, February 2 2008.

Yabuki N, Machinaka H, Li Z: 2006, Virtual reality with stereoscopic vision and Augmented Reality to steel bridge design and erection, *in*, *Proceedings of the Joint International Conference on Computing and Decision Making in Civil and Building Engineering*, Montreal, Canada, pp. 1284–1292.

Yang J, Yang W, Denecke M, Waibel A: 1999, Smart Sight: A tourist assistant system, *in*, *3rd International Symposium on Wearable Computers*, IEEE, San Francisco, CA, pp. 73–78.

Yeh M, Wickens CD, Seagull FJ: 1998, *Effects of Frame of Reference and Viewing Condition on Attentional Issues with Helmet Mounted Displays: Technical Report*, U.S. Army Research Laboratory, Interactive Displays Federated Laboratory, Aberdeen Proving Ground, MD.

Yeo W, Lou C, Kaga A, Sasada T, Byun C, Ikegami T: 2003, An interactive digital archive for Japanese historical architecture, *in*, *Proceedings of the 8th International Conference on Computer Aided Architectural Design Research in Asia (CAADRIA)*, Bangkok, Thailand, pp. 513–522.

Yip WH: 2001, *The Relative Functions of Text and Drawing in Computer-Supported Collaborative Problem-Solving*, M.Phil. thesis, Psychology, Cognitive Science, University of Hong Kong.

You S, Neumann U, Azuma R: 1999, Hybrid internal and vision tracking for augmented reality registration, *in*, *Proceedings of IEEE VR '99*, Houston, TX, pp. 260–267.

Young JO: 2001, *Art and Knowledge*, Routledge, London.

Zaki AR, Mailhot G: 2003, "Deck reconstruction of jacques cartier bridge using precast prestressed high performance concrete panels," *PCI Journal*, **48**(5), 20–33.

INDEX

M

N

O

P

W